SOLUTIONS MANUAL

for the

CHEMICAL ENGINEERING REFERENCE MANUAL

Randall N. Robinson, P.E.

PROFESSIONAL PUBLICATIONS, INC.
Belmont, CA 94002

In the ENGINEERING REVIEW MANUAL SERIES

Engineer-In-Training Review Manual
  Engineering Fundamentals Quick Reference Cards
  Mini-Exams for the E-I-T Exam
  1001 Solved Engineering Fundamentals Problems
  E-I-T Review: A Study Guide
Civil Engineering Reference Manual
  Civil Engineering Quick Reference Cards
  Civil Engineering Sample Examination
  Civil Engineering Review Course on Cassettes
  Seismic Design for the Civil P.E. Exam
  Timber Design for the Civil P.E. Exam
Structural Engineering Practice Problem Manual
Mechanical Engineering Review Manual
  Mechanical Engineering Quick Reference Cards
  Mechanical Engineering Sample Examination
  Mechanical Engineering Review Course on Cassettes
Electrical Engineering Review Manual
Chemical Engineering Reference Manual
  Chemical Engineering Practice Exam Set
Land Surveyor Reference Manual
Metallurgical Engineering Practice Problem Manual
Petroleum Engineering Practice Problem Manual
Expanded Interest Tables
Engineering Law, Design Liability, and Professional Ethics
Engineering Unit Conversions

In the ENGINEERING CAREER ADVANCEMENT SERIES

How to Become a Professional Engineer
The Expert Witness Handbook—A Guide for Engineers
Getting Started as a Consulting Engineer
Intellectual Property Protection—A Guide for Engineers
E-I-T/P.E. Course Coordinator's Handbook

Distributed by: Professional Publications, Inc.
1250 Fifth Avenue
Department 77
Belmont, CA 94002
(415) 593-9119

SOLUTIONS MANUAL for the
CHEMICAL ENGINEERING REFERENCE MANUAL

Printed in the United States of America

ISBN: 0-932276-78-4

Professional Publications, Inc.
1250 Fifth Avenue, Belmont, CA 94002

Current printing of this edition (last number): 6 5 4 3 2

TABLE OF CONTENTS

1 MATHEMATICS 1

2 STOICHIOMETRY, HEAT, AND MATERIAL BALANCES 5

3 ENGINEERING ECONOMICS 13

4 THERMODYNAMICS 15

5 FLUID STATICS AND DYNAMICS 21

6 HEAT TRANSFER: CONDUCTION AND RADIATION 27

7 HEAT TRANSFER: CONVECTION AND EQUIPMENT 29

8 VAPOR-LIQUID PROCESSES 35

9 DISTILLATION, EVAPORATION, AND HUMIDIFICATION 38

10 LIQUID-LIQUID AND SOLID-LIQUID PROCESSES 43

11 KINETICS 47

[1]

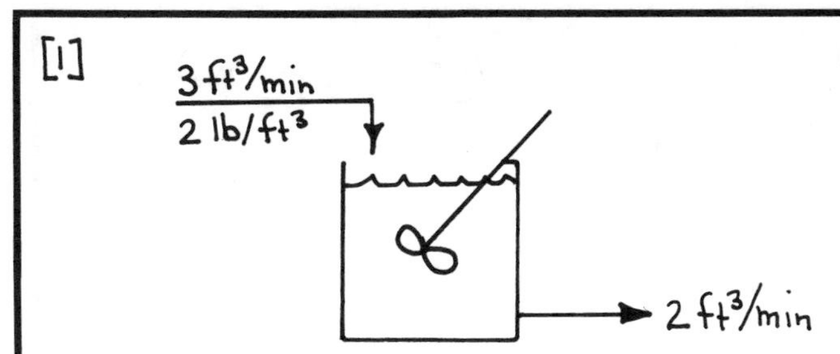

In - Out = Accum

$Q \equiv$ flow ft³/min : c_o = concentration out

t = time, min $V \equiv$ volume, ft³

c_i = concentration in

$$\therefore Q_i c_i - Q_o c_o = \frac{d}{dt}(Vc_o) = c_o\frac{dV}{dt} + V\frac{dc_o}{dt}$$

but $\frac{dV}{dt} = 1 \quad \therefore \int dV = \int dt$

or $V = t + k$ at $t = 0$, $V = 20$

$\therefore$ $k = 20$ and $V = t + 20$

$$(3)(2) - 2c_o = c_o(1) + [(1)t + 20]\frac{dc_o}{dt}$$

$$\int \frac{dc_o}{6 - 3c_o} = \int \frac{dt}{t + 20}$$

$$-\frac{1}{3}\ln(6 - 3c_o) = \ln(20 + t) + I$$

at $t = 0$ $c_o = 0$ $\therefore I = \ln\left[\frac{1}{20\sqrt[3]{6}}\right]$

plugging value for I and rearranging,

$$c_o = 2 - \frac{2}{(1 + .05t)^3}$$

when $V = 30$, $t = 10$

$$c_o = 2 - \frac{2}{(1 + .05(10))^3} = 1.407\ \frac{\text{lb}_m}{\text{ft}^3}$$

when $V = 50$, $t = 30$

$$c_o = 2 - \frac{2}{(1 + .05(30))^3} = 1.872\ \frac{\text{lb}_m}{\text{ft}^3}$$

but $1.872 < 2(.99) = 1.98$

Diff. eq. for overflow

$$(3)(2) - 3c_o = V\frac{dc_o}{dt} \qquad V = 50$$

$$\int \frac{dc_o}{6 - 3c_o} = \frac{1}{50}\int dt$$

$$-\frac{1}{3}\ln(6 - 3c_o) = \frac{1}{50}t + I$$

at $t = 0$ $c_o = 1.872$

$$I = \ln\left[\frac{1}{\sqrt[3]{.384}}\right]$$

$$\therefore t = 50\ln\left[\frac{1}{\sqrt[3]{(6 - 3c_o)(0.384)}}\right]$$

when $c_o = 1.98$

$$t = \frac{50}{3}\ln\left[\frac{1}{(.384)(6 - 3(1.98))}\right]$$

$t = 62.8$ min

total time = 30 + 62.8 = 92.8 min

[2]

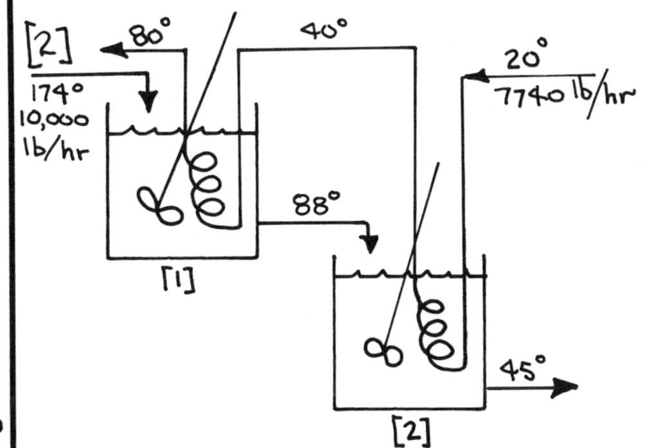

W_a = acid flow, lb/hr

C_a = acid specific heat = 0.36 Btu/lb°F

T_o = acid feed temp

T_1 = acid exit 1 temp °C

T_2 = acid exit 2 temp °C

V = tank cap., lb

t = time, hr

ASSUME

1. B.C. @ $t = 0$, $T_1 = 88°$; $T_2 = 45°$
2. Δt_m = log mean $\Delta t = (\Delta t_a - \Delta t_b)/\ln\frac{\Delta t_a}{\Delta t_b}$
3. area of coil = A ft³
4. heat capacity constant

Steady state conditions

Q_2 = heat removed in tank 2 (°F to °C)

$Q_2 = W_A C_A \Delta t = 10000(.36)(88-45)(1.8)$

$Q_2 = 278640$ Btu/hr

$$Q_2 = U_2 A_2 \Delta t_m = 130 A_2 \frac{48-25}{\ln 48/25}$$

$Q_2 = 4583.6\, A_2$

$A_2 = 278640/4583.6 = 60.79\ \text{ft}^2$

$Q_1 = W_a C_a \Delta t = 10000(.36)(174-88)(1.8)$

$Q_1 = 557280$ Btu/hr

$$Q_1 = U_1 A_1 \Delta t_m = 200 A_1 \frac{94-48}{\ln 94/48}$$

$Q_1 = 13688.56\, A_1$

$A_1 = 557280/13688.6 = 40.71\ \text{ft}^2$

[a]
Energy balance:

Tank 1:
$$W_a C_a T_0 - W_a C_a T_1 = \frac{d}{dt}(V_1 C_a T_1)$$

Tank 2:
$$W_a C_a T_1 - W_a C_a T_2 = \frac{d}{dt}(V_2 C_a T_2)$$

$$\therefore\ T_0 - T_1 = \frac{V_1}{W_a}\frac{dT_1}{dt}$$
$$T_1 - T_2 = \frac{V_2}{W_a}\frac{dT_2}{dt}$$

simultaneous 1st order diff. eq. solve tank 1 first

$$\int \frac{dT_1}{T_0 - T_1} = -\ln(T_0 - T_1) = \int \frac{W_a}{V_1}dt = \frac{W_a t}{V_1} + C$$

$$T_0 - T_1 = K e^{-W_a t/V_1}$$

B.C.: $t=0$ $T_1 = 88$ $\therefore K = 86$

$$T_0 - T_1 = 86\, e^{-W_a t/V} \quad \text{since } \frac{W_a}{V} = 1$$

at $t=1$ $T_1 = T_0 - 86e^{-1} = 142.4$°C after 1 hour cooling off.

tank 1: $T_1 = T_0 - 86 e^{W_a t/V_1}$

plug into tank 2 diff. eq.

$$T_0 - 86e^{-W_a t/V_1} - T_2 = \frac{V_2}{W_a}\frac{dT_2}{dt}$$

$$\frac{dT_2}{dt} + \frac{W_a}{V_2}T_2 = \frac{W_a}{V_2}\left[T_0 - 86e^{-W_a t/V_1}\right]$$

has form:

$$\frac{dy}{dx} + P(x)y = Q(x)$$

has solution

$$y = Ce^{-\int P dx} + e^{-\int P dx}\int e^{\int P dx} Q\, dx$$

$$\therefore\ T_2 = Ce^{-W_a t/V_2} + e^{-W_a t/V_2}\left[\int e^{W_a t/V_2}\left(\frac{W_a}{V_2}\right)\left(T_0 - 86e^{-W_a t/V_1}\right)dt\right]$$

$$\frac{W_a}{V_1} = \frac{W_a}{V_2} = 1$$

$$T_2 = Ce^{-t} + e^{-t}\left[\int e^{t}(T_0 - 86e^{-t})dt\right]$$

$$T_2 = Ce^{-t} + T_0 - 86\,t\,e^{-t}$$

@ $t=0$ $T_2 = 45$

$45 = C + 174$ $\therefore C = -129$

$$\therefore\ T_2 = T_0 - (86t + 129)e^{t}$$

at $t=1$

$T_2 = 174 - (86 + 129)e$

$T_2 = 94.9$ °C

[b] let new water W lb/hr

T_3 = water exit tank 2

T_4 = water exit tank 1

C_w = water heat capacity

T_c = water inlet temp

Energy balance (acid & water)

Tank 1: $W c_w T_4 + W_a C_a T_0 - (W c_w T_3 + W_a C_a T_1) = V_1 C_a \dfrac{dT_1}{dt}$

Tank 2:

$$W c_w T_c + W_a C_a T_1 - (W c_w T_4 + W_a C_a T_2) = V_1 C_a \frac{dT_2}{dt}$$

[2] con't

Heat transfer tank 2:

$$Wc_w(T_3-T_c) = U_2A_2 \frac{(T_1-T_3)-(T_2-T_c)}{\ln\frac{(T_1-T_3)}{(T_2-T_c)}}$$

Heat transfer tank 1:

$$Wc_w(T_4-T_3) = U_1A_1 \frac{(T_0-T_4)-(T_1-T_3)}{\ln\frac{T_0-T_4}{T_1-T_3}}$$

to simplify,

let $\alpha = e^{-U_1A_1/Wc_w}$

$\beta = e^{-U_2A_2/Wc_w}$

$$\therefore \frac{T_2-T_3}{T_2-T_c} = \beta\,; \quad \frac{T_1-T_4}{T_1-T_3} = \alpha$$

or $T_2(1-\beta) = T_3-\beta T_c$; $T_1(1-\alpha) = T_4-\alpha T_3$

4 simultaneous equations result.

$$(1)\quad T_4+C_aT_0-(T_3+C_aT_1) = C_a\frac{dT_1}{dt}$$

$$(2)\quad T_c+C_aT_1-(T_4+C_aT_2) = C_a\frac{dT_2}{dt}$$

$$(3)\quad T_2(1-\beta) = T_3-\beta T_c$$

$$(4)\quad T_1(1-\alpha) = T_4-\alpha T_3$$

ELIMINATE T_4 by substituting (4) into (2) & (1) and ELIMINATE T_3 by substituting (3) into (1) & (2) results:

$$(1-\alpha)(1-\beta)T_2+(1-\alpha)\beta T_c+C_a(T_0-T_1)-(1-\alpha)T_1 = C_a\frac{dT_1}{dt} \qquad (5)$$

$$T_c+C_a(T_1-T_2)-(1-\beta)T_2-\beta T_c = C_a\frac{dT_2}{dt} \qquad (6)$$

differentiating above equation

$$C_a\frac{dT_1}{dt} = C_a\frac{d^2T_2}{dt^2}+(1-\beta+C_A)\frac{dT_2}{dt} \qquad (7)$$

using equation (6) to eliminate T_1 in (7) results in (8):

$$C_a^2\frac{d^2T_2}{dt^2}+(2C_a+2-\alpha-\beta)C_a\frac{dT_2}{dt}+$$

$$(C_a^2+C_a(1-\alpha\beta)+(1-\alpha)(1-\beta))T_2$$

$$= [C_a(1-\alpha\beta)+(1-\alpha)(1-\beta)]T_c+C_A^2T_0$$

$$\frac{U_1A_1}{W} = \frac{200(40.71)}{10\,000} = .8141\,; \quad \alpha = 0.44299$$

$$\frac{U_2A_2}{W} = \frac{130(60.79)}{10\,000} = .79027\,; \quad \beta = 0.4537$$

Eq. (8) becomes

$$\frac{d^2T_2}{dt^2}+6.06\frac{dT_2}{dt}+7.7\,T_2 = 308$$

which is 2nd order diff eq. with constant coefficients

$$\therefore T_2 = Ae^{-4.25t}+Be^{-1.81t}+40$$

from B.C. $A = 0.6$; $B = 54.3$

$$T_2 = 0.6e^{-4.25t}+54.3e^{-1.81t}+40$$

at $t = 1$ hour

$$T_2 = 0.6e^{-4.25}+54.3e^{-1.81}+40$$

$$T_2 = 48.9°C$$

[3] Use 3-point method

form: $y = a+bx+c/x^2$

$$b = \frac{f\sigma-pes^2t^2}{fm-nes^2t^2}$$

using last 3 data points,

$m = x_2-x_1 = 180-100 = 80$

$n = x_3-x_2 = 320-180 = 140$

$\sigma = y_2-y_1 = 1.3996-1.3779 = 0.0217$

$p = y_3-y_2 = 1.4391-1.3996 = 0.0395$

[3] con't

Since T must be in °R

$$s = \frac{X_2}{X_1} = \frac{180+460}{100+460} = 1.1429$$

$$t = \frac{X_3}{X_2} = \frac{320+460}{180+460} = 1.2188$$

$$f = X_3^2 - X_2^2 = 780^2 - 640^2 = 198800.$$

$$e = X_2^2 - X_1^2 = 640^2 - 560^2 = 96000.$$

$$b = \frac{(198800)(.0217) - .0395(9600)[1.1429 \times 1.2188]^2}{(198800)(80) - 140(9600)[1.1429 \times 1.2188]^2}$$

$$b = 2.691 \times 10^{-4}$$

$$c = \frac{(bm - \sigma)x_1^4}{es^2}$$

$$c = \frac{(2.691 \times 10^{-4}(80) - .0217)(560)^4}{(9600)(1.1429)^2}$$

$$c = -1348.9$$

$$a = y_1 - bx_1 - c/x_1^2$$

$$a = 1.3779 - 2.691 \times 10^{-4}(560) - \frac{-1348.9}{560^2}$$

$$a = 1.2315$$

for $t = 250°F$; $T = 250 + 460 = 710°R$

$$C_p = 1.2315 + 2.691 \times 10^{-4}(710) - \frac{1348.9}{710^2}$$

$$C_p = 1.4199$$

[4] $kt = \frac{1}{C_a} - \frac{1}{C_{a_o}}$

let $\frac{1}{C_a} = y$; $\frac{1}{C_{a_o}} = b$; $t = x$; $k = m$

$y = mx + b$ linear

C_a	y	x	xy	x^2
2.2502	0.4444	2	0.8888	4
1.5888	0.6294	7	4.4058	49
1.4217	0.7034	9	6.3304	81
1.1745	0.8514	13	11.0685	169
0.5298	1.8875	41	77.3877	1681
0.3753	2.6645	62	165.2012	3844
	7.1807	134	265.2825	5828

$$m = \frac{n\Sigma xy - \Sigma x \Sigma y}{n\Sigma x^2 - (\Sigma x)^2}$$

$$m = \frac{6(265.2825) - 134(7.1097)}{6(5828) - 134^2}$$

$m = k = 0.0376$ liter/mole-min

$$b = \bar{y} - m\bar{x}$$

$$b = \frac{7.1807}{6} - .0376\frac{134}{6}$$

$$b = .3571$$

$C_{a_o} = 1/b = 1/.3571 = 2.800$ mole/liter

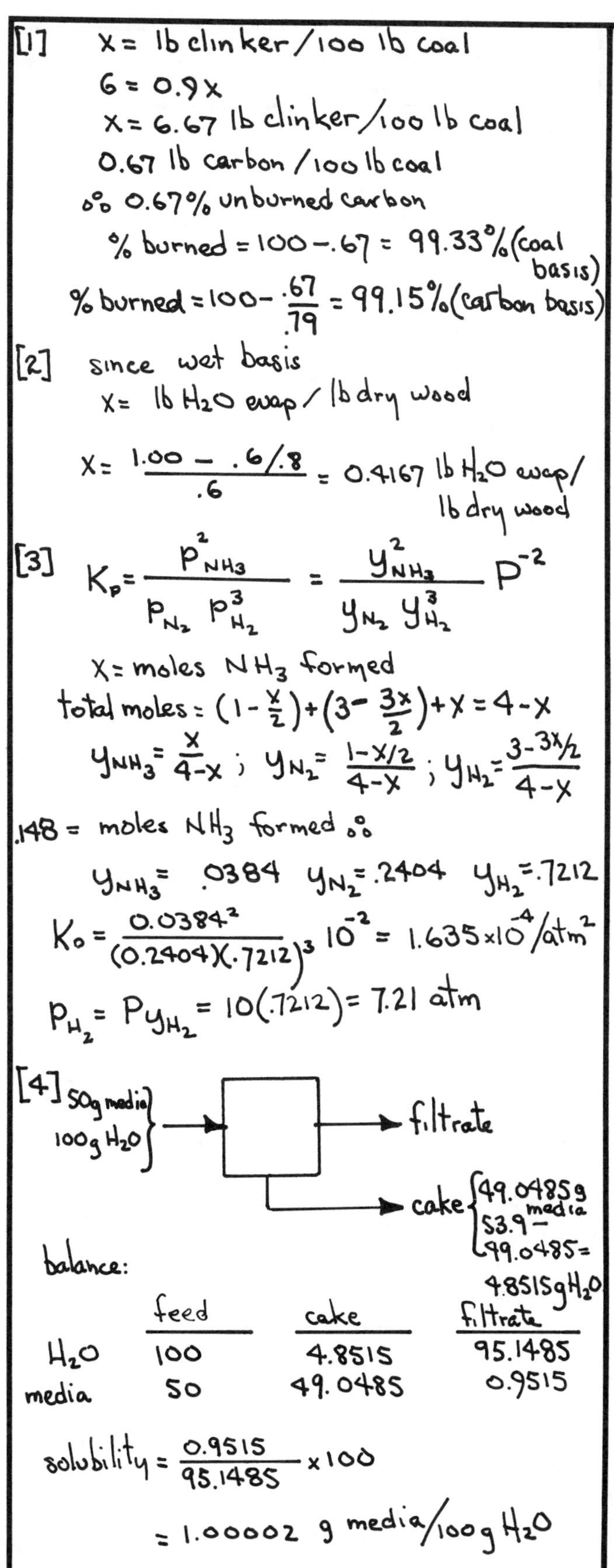

[1] x = lb clinker/100 lb coal

$6 = 0.9x$

$x = 6.67$ lb clinker/100 lb coal

0.67 lb carbon/100 lb coal

∴ 0.67% unburned carbon

% burned = 100 − .67 = 99.33% (coal basis)

% burned = $100 - \frac{.67}{.79}$ = 99.15% (carbon basis)

[2] since wet basis

x = lb H_2O evap/lb dry wood

$$x = \frac{1.00 - .6/.8}{.6} = 0.4167 \text{ lb } H_2O \text{ evap/lb dry wood}$$

[3] $$K_p = \frac{p_{NH_3}^2}{p_{N_2}\, p_{H_2}^3} = \frac{y_{NH_3}^2}{y_{N_2}\, y_{H_2}^3} P^{-2}$$

x = moles NH_3 formed

total moles = $(1 - \frac{x}{2}) + (3 - \frac{3x}{2}) + x = 4 - x$

$y_{NH_3} = \frac{x}{4-x}$; $y_{N_2} = \frac{1 - x/2}{4-x}$; $y_{H_2} = \frac{3 - 3x/2}{4-x}$

.148 = moles NH_3 formed ∴

$y_{NH_3} = .0384$ $y_{N_2} = .2404$ $y_{H_2} = .7212$

$$K_o = \frac{0.0384^2}{(0.2404)(.7212)^3} 10^{-2} = 1.635\times10^{-4}/\text{atm}^2$$

$p_{H_2} = P y_{H_2} = 10(.7212) = 7.21$ atm

[4] 50 g media, 100 g H_2O → filtrate; → cake {49.0485 g media; 53.9 − 49.0485 = 4.8515 g H_2O}

balance:

	feed	cake	filtrate
H_2O	100	4.8515	95.1485
media	50	49.0485	0.9515

$$\text{solubility} = \frac{0.9515}{95.1485} \times 100$$

$= 1.00002$ g media/100 g H_2O

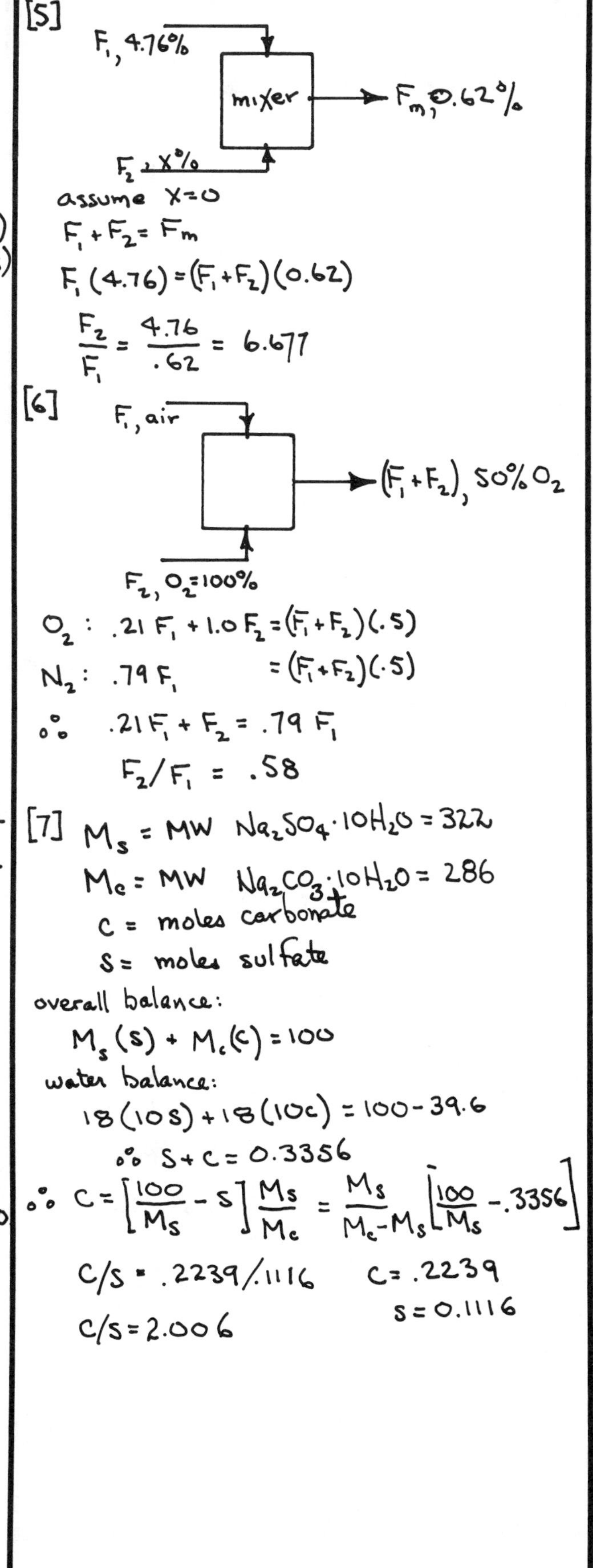

[5] F_1, 4.76% → mixer → F_m, 0.62%; F_2, x%

assume $x = 0$

$F_1 + F_2 = F_m$

$F_1(4.76) = (F_1 + F_2)(0.62)$

$$\frac{F_2}{F_1} = \frac{4.76}{.62} = 6.677$$

[6] F_1, air; F_2, O_2 = 100% → $(F_1 + F_2)$, 50% O_2

O_2: $.21F_1 + 1.0F_2 = (F_1 + F_2)(.5)$

N_2: $.79F_1 = (F_1 + F_2)(.5)$

∴ $.21F_1 + F_2 = .79F_1$

$F_2/F_1 = .58$

[7] M_s = MW $Na_2SO_4 \cdot 10H_2O$ = 322

M_c = MW $Na_2CO_3 \cdot 10H_2O$ = 286

c = moles carbonate

s = moles sulfate

overall balance:

$M_s(s) + M_c(c) = 100$

water balance:

$18(10s) + 18(10c) = 100 - 39.6$

∴ $s + c = 0.3356$

$$\therefore c = \left[\frac{100}{M_s} - s\right]\frac{M_s}{M_c} = \frac{M_s}{M_c - M_s}\left[\frac{100}{M_s} - .3356\right]$$

$c/s = .2239/.1116$ $c = .2239$

$c/s = 2.006$ $s = 0.1116$

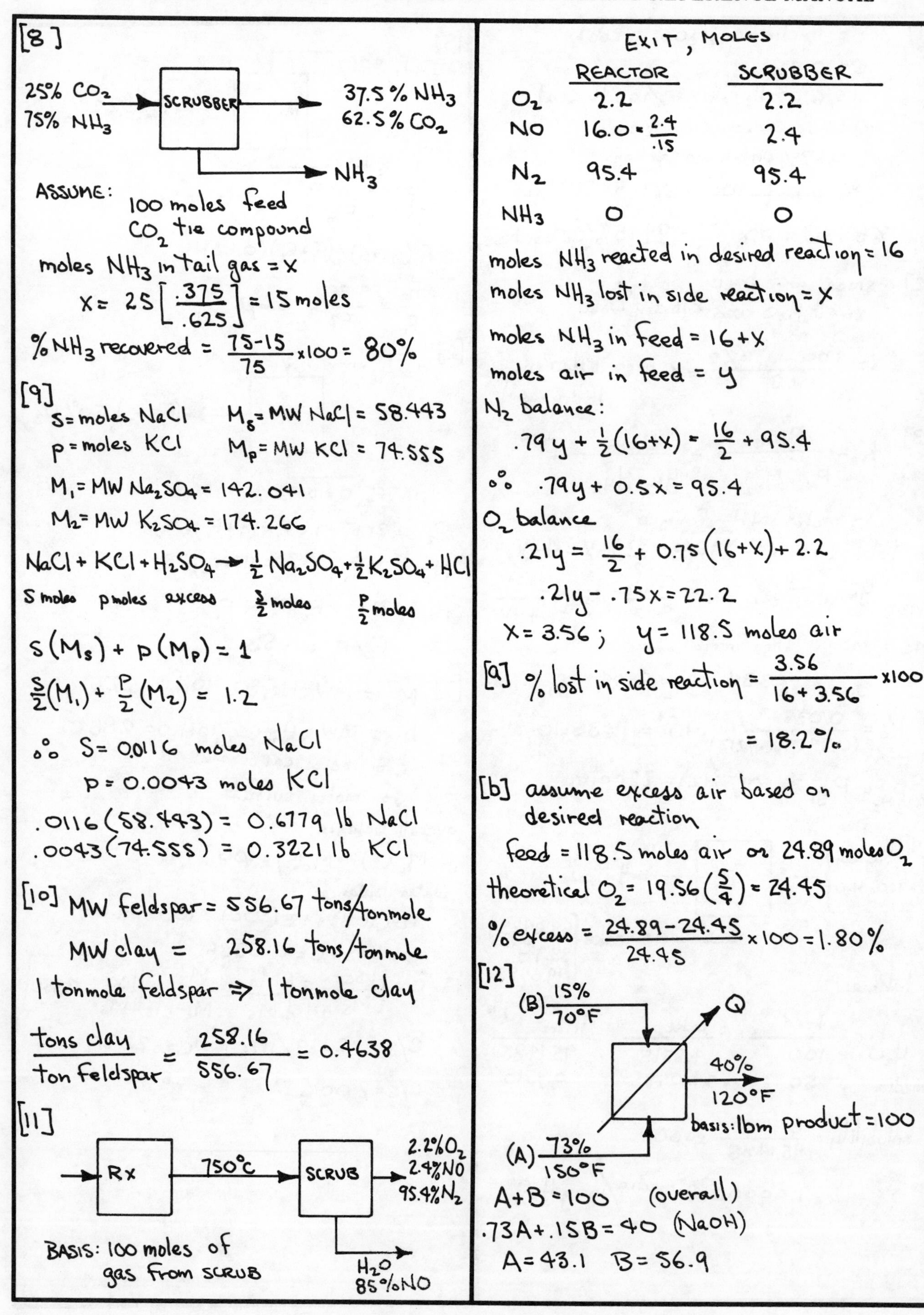

[8]

25% CO_2, 75% NH_3 → SCRUBBER → 37.5% NH_3, 62.5% CO_2; SCRUBBER → NH_3

ASSUME: 100 moles feed
CO_2 tie compound

moles NH_3 in tail gas = x

$$x = 25\left[\frac{.375}{.625}\right] = 15 \text{ moles}$$

$$\%NH_3 \text{ recovered} = \frac{75-15}{75} \times 100 = 80\%$$

[9]

s = moles NaCl M_s = MW NaCl = 58.443
p = moles KCl M_p = MW KCl = 74.555
M_1 = MW Na_2SO_4 = 142.041
M_2 = MW K_2SO_4 = 174.266

$$NaCl + KCl + H_2SO_4 \rightarrow \tfrac{1}{2} Na_2SO_4 + \tfrac{1}{2} K_2SO_4 + HCl$$

s moles p moles excess $\frac{s}{2}$ moles $\frac{p}{2}$ moles

$$s(M_s) + p(M_p) = 1$$

$$\frac{s}{2}(M_1) + \frac{p}{2}(M_2) = 1.2$$

∴ s = 0.0116 moles NaCl
p = 0.0043 moles KCl

.0116(58.443) = 0.6779 lb NaCl
.0043(74.555) = 0.3221 lb KCl

[10] MW feldspar = 556.67 tons/tonmole
MW clay = 258.16 tons/tonmole
1 tonmole feldspar ⇒ 1 tonmole clay

$$\frac{\text{tons clay}}{\text{ton feldspar}} = \frac{258.16}{556.67} = 0.4638$$

[11]

→ Rx → 750°C → SCRUB → 2.2% O_2, 2.4% NO, 95.4% N_2; SCRUB → H_2O, 85% NO

BASIS: 100 moles of gas from SCRUB

EXIT, MOLES

	REACTOR	SCRUBBER
O_2	2.2	2.2
NO	$16.0 = \frac{2.4}{.15}$	2.4
N_2	95.4	95.4
NH_3	0	0

moles NH_3 reacted in desired reaction = 16
moles NH_3 lost in side reaction = x
moles NH_3 in feed = 16 + x
moles air in feed = y

N_2 balance:

$$.79y + \tfrac{1}{2}(16+x) = \frac{16}{2} + 95.4$$

$$\therefore .79y + 0.5x = 95.4$$

O_2 balance

$$.21y = \frac{16}{2} + 0.75(16+x) + 2.2$$

$$.21y - .75x = 22.2$$

x = 3.56; y = 118.5 moles air

[a] $$\%\text{ lost in side reaction} = \frac{3.56}{16+3.56} \times 100 = 18.2\%$$

[b] assume excess air based on desired reaction

feed = 118.5 moles air or 24.89 moles O_2

$$\text{theoretical } O_2 = 19.56\left(\frac{5}{4}\right) = 24.45$$

$$\%\text{ excess} = \frac{24.89-24.45}{24.45} \times 100 = 1.80\%$$

[12]

(B) 15%, 70°F → ; (A) 73%, 150°F → ; → 40%, 120°F; Q

basis: lbm product = 100

A + B = 100 (overall)
.73A + .15B = 40 (NaOH)
A = 43.1 B = 56.9

heat balance:

$$340A + 30B = 107(40) + Q$$

$$Q = 340(43.1) + 30(56.9) - 107(40)$$

$$Q = 5661 \text{ Btu/100 lbm product}$$

[13] MW

NaCN = 49.02
NaOH = 40
Cl_2 = 70.91
NaCNO = 65.02

$$\text{NaCN/day} = 3600\frac{\text{gal}}{\text{day}}\left(8.33\frac{\text{lb}}{\text{gal}}\right)(0.0164) = 492\frac{\text{lb}}{\text{day}}$$

$$\text{NaCNO/day} = 492\left(\frac{65.02}{49.02}\right) = 652\frac{\text{lb}}{\text{day}}$$

Caustic: 1st step

$$\text{lb NaOH} = 1.15(492)\left[\frac{2(40)}{49.02}\right] = 922\frac{\text{lb}}{\text{day}}$$

Caustic: 2nd step

$$\text{lb NaOH} = 1.15(652)\left[\frac{4(40)}{2(65.02)}\right] = 922\frac{\text{lb}}{\text{day}}$$

1st step: 922 = 802 + 120 excess

2nd step: 922 = 802 + 120 excess

[a] use same excess for both steps

total NaOH needed = 802 + 802 + 120
= 1724 $\frac{\text{lb NaOH}}{\text{day}}$

Cl_2: 1st step

$$\text{lb}Cl_2 = 1.20(492)\left[\frac{70.91}{49.02}\right] = 853 \text{ lb}$$

(711 lb + 142 lb excess)

Cl_2 2nd step

$$\text{lb}Cl_2 = 1.20(652)\left[\frac{3(70.91)}{2(65.02)}\right] = 1278 \text{ lb}$$

(1065 lb + 213 lb excess)

use 213 lb excess for both steps

$$\text{lb}Cl_2 = 711 + 1065 + 213 = 1989\frac{\text{lb } Cl_2}{\text{day}}$$

$$\text{NaOH min weight \%} = \frac{1724}{(9600)(8.33)} \times 100$$

$$= 2.16\%$$

[b] heat balance: $mc_p\Delta t$

$$3600(1)(90°F - 85°F) + 9600(1)(120° - 85°) = X(1)(85° - 75°)$$

$X = \frac{\text{gal}}{\text{day}}$ cooling water

$$X = 35400\frac{\text{gal}}{\text{day}}$$

[c] NaCl formed

$$1^{st}\text{ step: } 492\left(\frac{116.91}{49.02}\right) = 1173 \text{ lb}$$

$$2^{nd}\text{ step } 652\left(\frac{6(58.45)}{2(65.02)}\right) = 1758 \text{ lb}$$

total NaCl formed = 1173 + 1758
= 2931 lb/day

$$\text{NaCl} = 10^6\frac{2931}{48600(8.33)} = 7300 \text{ ppm}$$

[14]

16.0% C_2H_4
19.9% CH_4
32.3% H_2
26.1% CO_2
2.9% CO
2.8% N_2

air

H_2O

flue gas (f.g.)
11.83 % CO_2
0.4 % CO
4.53% O_2
83.44% N_2

BASIS 100 lbmoles feed

ELEMENTS : FEED, lb MOLES

	C	O_2	H_2	N_2
C_2H_4	16		32	
CH_4	19.9		39.8	
H_2			32.3	
CO_2	26.1	26.1		
CO	2.9	1.45		
N_2				2.8
	64.9	27.55	104.1	2.8

moles H_2O in f.g. = moles H_2 in feed

$= 104.1 \frac{\text{moles } H_2O \text{ in f.g}}{100 \text{ moles feed}}$

Carbon used to determine moles dry f.g.

$$\text{moles dry f.g} = \frac{64.9}{(.1183+.004)} = 530.2$$

moles wet f.g. = 530.2 + 104.1

$= 634.3 \frac{\text{moles f.g}}{100 \text{ moles feed}}$

Air from N_2 balance: x = lbmoles air

$.79x + 2.8 = .8344(530.2)$

$x = 556.5$ lbmoles air

[a] since mole ratio = volume ratio

air = 5.565 ft^3 air/ft^3 feed

[b] at 670°F (1130°R)

and feed at 68°F (528°R)

$$\text{flue gas} = 6.343\left[\frac{1130}{528}\right]$$

$= 13.575 \frac{ft^3 \text{ f.g.}}{ft^3 @ 68°F}$

[15] M_S = MW Na_2SO_4 = 142.04

M_W = MW H_2O = 18.01

M_D = MW $Na_2SO_4 \cdot 10H_2O$ = 322.14

X_S = moles $NaSO_4$ in solid

$$\text{solubility} = \frac{\frac{10}{M_s} - X_s}{\frac{90}{M_W} - 10X_s} = .00634$$

since we want $X_S M_D$

$$X_S M_D = M_D\left[\frac{10}{M_S} - \frac{90}{M_W}(.00634)\right]/.9366$$

$$X_S M_D = 322.14\left[\frac{1.0}{142.04} - \frac{90}{18.01}(.00634)\right]/.9366$$

$X_S M_D$ = 13.318 lb solid/100 lb mix

[16] feed, lb moles 100 lbmole basis

	C	H_2
C_4H_{16}	350	800
C_8H_{18}	400	900
	750	1700

$$O_2 \text{ req'd} = 1.1\left[750 + \tfrac{1}{2}(1700)\right]$$

$= 1760$ lb mole O_2/100 lbmole fuel

air = 1760/.21 = $8380.9 \frac{\text{lb mole air}}{100 \text{ lb mole fuel}}$

100 lb moles fuel = 50(100 + 114) = 10700 lb fuel

$$\text{air} = 8390.9\left[\frac{100}{10700}\right]$$

= 78.41 lb moles air/100 lb fuel

[a]

$$\frac{ft^3 \text{ air}}{100 \text{ lb fuel}} = 78.41\left[359 \frac{ft^3}{\text{lbmole}}\right]\left[\frac{530°R}{492°R}\right]\left[\frac{14.7 \text{ psi}}{14.9 \text{ psi}}\right]$$

$= 29919$ ft^3 air/100 lb fuel

basis 100 lbmole fuel

CO_2 750

H_2O 1700

O_2 $1760 - 750 - \frac{1}{2}(1700) = 160$

N_2 $8380.9(.79) = 6620.9$

total = 9230.9

[b]

CO_2 = 750/9230.9 = 8.1%

H_2O = 1700/9230.9 = 18.4%

O_2 = 160/9230.9 = 1.73%; N_2 = 71.8%

[17]

H_2O: 58.36%
$Ca(OCl)_2$: 27.77% → W_s → filter → W_f → H_2O: 69.60%, $Ca(OCl)_2$: 12.92%

filter → W_c → H_2O: 47.5%, $Ca(OCl)_2$: 42.35%

Assume: (1) remaining % impurities
(2) precipitate pure $Ca(OCl)_2 \cdot 2H_2O$
(3) 100 lb feed

H_2O balance:

$$58.36 = .696\,W_f + .475\,W_c$$

$$W_f + W_c = 100$$

$\therefore$ $W_f = 49.14$ lb ; $W_c = 50.86$ lb

[a] slurry consists of pure $Ca(OCl)_2$ + mother liquor

$$W_{m.l.} + W_{SOLID} = 100$$

H_2O balance

$$.7128\,W_{m.l} + \frac{36.04}{179.02} W_{SOLID} = 58.36$$

$W_{m.l} = 74.74$ lb ; $W_{SOLID} = 25.26$ lb

$\therefore$ slurry = 25.26% solid

[b] $\dfrac{\text{lb solid lost}}{\text{100 lb filtrate}} =$

$$\left[12.92 - \frac{10.2 \text{ lb } Ca(OCl)_2 \text{ dissolved}}{71.28 \text{ lb } H_2O} \times 69.2 \text{ lb } H_2O\right]$$

$$\times \frac{179.02}{142.98} = 3.71 \frac{\text{lb solid lost}}{\text{100 lb filtrate}}$$

3.71% solid in filtrate

[c]

$$\frac{3.71 \text{ lb solid}}{100 \text{ lb filtrate}} \times \frac{49.14 \text{ lb filtrate}}{100 \text{ lb slurry}} \times \frac{100 \text{ lb slurry}}{26.26 \text{ lb solid}}$$

$$= 7.22 \frac{\text{lb solid lost}}{\text{100 lb solid in slurry}}$$

% lost = 7.22%

[18]

$$C_3H_8 + (5+15)O_2 + (5+15)\frac{.79}{.21}N_2 \rightarrow$$

$$3CO_2 + 4H_2O + 15O_2 + 20\frac{.79}{.21}N_2$$

basis: 1 mole fuel = 44.092 lb

product:

3 lbmoles CO_2: 84 lb
15 lbmoles O_2: 240 lb
$20\frac{.79}{.21}$ lbmoles N_2: 1053 lb
4 lbmoles H_2O: 64 lb

LHV = 19,944 Btu/lb

since hot air is being used the added sensible heat must be added to LHV to find TFT

assume TFT = 2000°F

	m	C_p	mC_p
CO_2	84	.275	23.1
H_2O	64	.515	32.96
O_2	240	.25	60.
N_2	1053	.27	284.31
			400.37

lb air used (300% excess) = 1053 + 20(16) = 1373 lb air

C_p air = .241

$\Delta t = 260 - 77 = 183$°F

sensible heat from air = 1373(.241)(183) = 1556 Btu

$$TFT = \frac{(44.092)(19{,}944) + 1556}{400.37} + 77$$

TFT = 2277°F

Assume TFT = 2200°F

	m	C_p	mC_p
CO_2	84	.28	23.52
H_2O	64	.520	33.28
O_2	240	.252	60.48
N_2	1053	.275	289.58

[18] con't

$$\Sigma m c_p = 406.9$$

$$TFT = \frac{(44.092)(19944)}{406.9} + 77$$

$$TFT = 2242°F$$

$$TFT = \frac{2277 + 2242}{2} = 2260°F$$

[19]

h = height, inches; c = cost, \$

d = diameter, inches

$$c = \frac{2\pi d^2 (7.00)}{4\ (144)} + \frac{25 t \pi d h}{144}$$

$$c = 0.07635 d^2 + 0.5454\, dht$$

$$t = Pd/2400$$

$$P = \rho h = \frac{2\ (62.4)}{1728} h = 0.07222 h$$

$$\therefore t = 3.009 \times 10^{-6}\, dh$$

$$h = \frac{20000\ (231)}{\pi d^2 / 4} = 5.882 \times 10^6 / d^2$$

$$\therefore C = 0.07635 d^2 + 5.678 \times 10^7 / d^2$$

find minimum, differentiate, set to 0

$$\frac{d}{dd}(c) = 0 = 0.1527 d - 1.1356 \times 10^8 / d^3$$

d = 215.5 inches

t = 0.107 too thin (t = .25)

with t = .25

$$C = 0.07635 d^2 + 8.02 \times 10^5 / d$$

$$\frac{d}{dd}(c) = 0 = 0.1527 d - 8.02 \times 10^5 / d^2$$

d = 173.9 inches or 14.5 ft

$$h = 5.882 \times 10^6 / d^2 = 194.5 \text{ inches}$$

or 16.2 ft.

[20]

$$CaCO_3 \rightarrow CO_2 + CaO \quad (1)$$

$$CH_4 + 2O_2 \rightarrow CO_2 + 2H_2O \quad (2)$$

$$CH_4 + \tfrac{3}{2} O_2 \rightarrow CO + 2H_2O \quad (3)$$

assume: missing % gas is O_2

100 lbmole f.g.

flue gas:

	moles	moles O_2	moles C
CO_2	20.4	20.4	20.4
CO	0.4	0.2	0.4
O_2	2.1	2.1	
N_2	77.1		
	100.0	22.7	20.8

$$O_2 \text{ from air} = 77.1\left(\frac{.79}{.21}\right) = 20.5 \ \frac{\text{lbmole } O_2}{100 \text{ lbmole f.g.}}$$

from equation (3)

$$CH_4 \text{ converted to } CO = 0.4 \ \frac{\text{lbmoles } CH_4}{100 \text{ lbmoles f.g.}}$$

$$O_2 \text{ consumed in (3)} = 0.6 \ \frac{\text{lbmoles } O_2}{100 \text{ lbmoles f.g.}}$$

from equation (2)

$$O_2 \text{ available} = 20.5 - 2.1 - 0.6 = 17.8 \ \frac{\text{lbmoles } O_2}{100 \text{ lbmoles f.g.}}$$

$$CH_4 \text{ burned to } CO_2 = \frac{17.8}{2} = 8.9 \ \frac{\text{lbmoles } CH_4}{100 \text{ lbmoles f.g.}}$$

$$\text{total } CH_4 \text{ burned} = 8.9 + 0.4 = 9.3 \ \frac{\text{lbmoles } CH_4}{100 \text{ lbmoles f.g.}}$$

$$CO_2 \text{ produced from calcine} = 20.4 - 8.9 = 11.5 \ \frac{\text{lbmoles } CO_2}{100 \text{ lbmoles f.g.}}$$

equation (1)

$$CaO \text{ produced} = 11.5 \ \frac{\text{lbmole CaO}}{100 \text{ lbmole f.g.}}$$

$$\text{Actual } CH_4 \text{ burned} = \frac{29000 \text{ ft}^3}{359 \text{ ft}^3/\text{lbmole}} \times \frac{492°R}{520°R} = 76.43 \text{ lbmole/hr}$$

$$\text{moles f.g. produced} = 76.43 \frac{100}{9.3} = 821.8 \text{ lbmoles f.g./hr}$$

$$CaO \text{ produced} = (11.5/100) \times 821.8 \times 56 = 5293 \text{ lb CaO/hr}$$

[21]

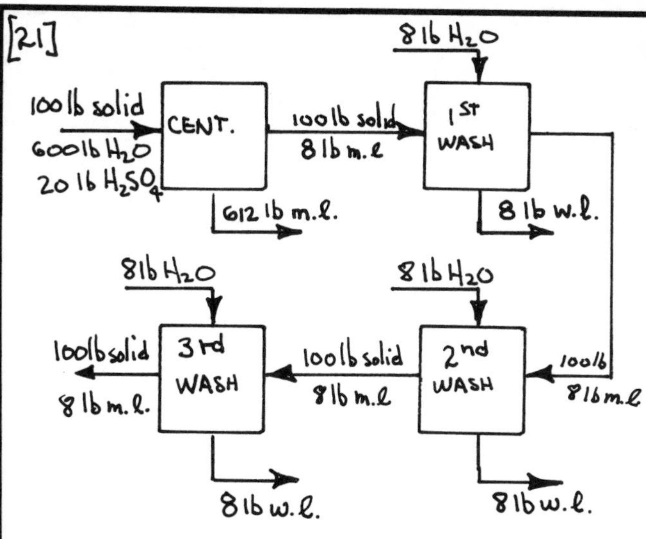

[a] unwashed cake:

mother liquor acid $= \frac{20}{620}(8)$

1st wash: $= 0.2581$ lb H_2SO_4

% acid in cake (dry basis) =

$$\frac{(0.2)(.2581)}{(100+(.2)(.2581))} \times 100 = 0.0516\%$$

[b] acid in 2nd wash $= (0.2)(0.2)(.2581)$
$= .010324$ lb

% acid in 2nd wash cake =

$$\frac{.010324}{(100+.01324)} \times 100 = 0.0132\%$$

2 stages needed

[c] acidity $= .0132\%$

total acid in wash liquor =

$0.2581 - .01324 = .24776$ lb H_2SO_4

total combined wash liquor = 16 lb

$$\% \text{ acid} = \frac{.24776}{16} \times 100 = 1.549\% \ H_2SO_4$$

[22] In − Out = Accum

$$10(.22) - 5c_o = \frac{d}{dt}(Vc_o)$$

$$2.2 - 5c_o = V\frac{dc_o}{dt} + c_o\frac{dV}{dt}$$

$$\frac{dV}{dt} = 10 - 5 = 5 \text{ gpm}$$

$$V = 5000 + 5t$$

$$\therefore 2.2 - 5c_o = (5000+5t)\frac{dc_o}{dt} + 5c_o$$

$$\int \frac{dc_o}{2.2 - 10c_o} = \int \frac{dt}{5000+5t}$$

$$-10\ln(2.2 - 10c_o) = 5\ln(5000+5t) + I$$

when $t = 0$ $c_o = 0.05$

$I = 47.892$

rearranging, dividing by 10, raising e^x

$$10c_o = 2.2 - \frac{120.20}{\sqrt{5000+5t}}$$

[a] when $t = 5 \times 60 = 300$ min

$$10c_o = 2.2 - \frac{120.20}{\sqrt{5000+5(300)}}$$

$c_o = 0.07091$ or 7.091%

[b] if overflow occurs at 10,000 gal then after 12 hours

total volume $= 5000 + 5(12 \times 60)$
$= 8600$ gal, no overflow

$$\therefore 10c_o = 2.2 - \frac{120.2}{\sqrt{8600}}$$

$c_o = 0.09038$ or 9.038%

[23]

Since no other term except C_1 is affected by the recycle stream:

$$C_2 = C_1^* (1-q) = \left[\frac{C_1 + C_2 R}{1+R}\right](1-q)$$

rearranging to make explicit in C_2

$$C_2 = \frac{C_1 (1-q)}{[1+R-R(1-q)]}$$

for $R=0$, $q = 2/3$

$$C_2 = \frac{C_1 (1-2/3)}{[1+0+0]} = \frac{1}{3} C_1$$

for $R=3$, $q = 2/3$

$$C_2 = \frac{C_1 (1-2/3)}{[1+3-3(1-2/3)]} = \frac{1}{9} C_1$$

removal efficiency before recycle =

$$100\left(1-\frac{1}{3}\right) = 66.7\%$$

after recycle starts =

$$100\left(1-\frac{1}{9}\right) = 88.9\%$$

$$\%\ \text{increase} = \frac{88.9-66.7}{66.7} \times 100 = 33\%$$

increase in removal efficiency.

at 95% removal

$$C_2 = \frac{1}{20} C_1$$

$$\therefore \frac{C_1 (1-q)}{[1+R-R(1-q)]} = \frac{1}{20} C_1$$

$q = 2/3$

$$\frac{1/3}{[1+R-R/3]} = \frac{1}{3+2R} = \frac{1}{20}$$

$$R = 17/2 = 8.5$$

A recycle ratio of 8.5 will remove 95% of rubidium with a removal fraction of 2/3.

[1] $F = P(F/P, 6\%, 10)$
$F = 1000(1.7908) = \$1790.80$

[2] $P = F(P/F, 6\%, 4)$
$P = 2000(0.7921) = \$1584.20$

[3] $P = F(P/F, 6\%, 20)$
$P = 2000(0.3118) = \$623.60$

[4] $A = P(A/P, 6\%, 7)$
$A = 500(.1791) = \$89.55/\text{year}$

[5] $F = A(F/A, 6\%, 10)$
$F = 50(13.1808) = \$659.04$

[6] if each year is independent
$P = F(P/F, 6\%, 1)$
$P = 200(.9434) = \$188.68$

[7] $A = F(A/F, 6\%, 5)$
$A = 2000(0.1774) = \$354.80/\text{year}$

[8] $F = P[(F/P, 6\%, 10) + (F/P, 6\%, 8) + (F/P, 6\%, 6)]$
$F = 100(1.7908 + 1.5938 + 1.4185)$
$F = \$480.13$

[9] $\phi = r/K \qquad n = 5(12) = 60$
$\phi = .06/12 = .005$ or 0.5%
$F = P(F/P, 0.5\%, 60)$
$F = 500(1.3489) = \$674.45$

[10] $P = F(P/F, i\%, 7)$
$80 = 120(P/F, i\%, 7)$
$(P/F, i\%, 7) = 0.6666$
$i \approx 6\%$ from tables

[11] $EUAC = (17000 + 5000)(A/P, 6\%, 5) + 200 - (14000 + 2500)(A/F, 6\%, 5)$
$EUAC = (22000)(.2374) + 200 - (16500)(0.1774)$
$= \$2495.70/\text{year}$

[12] ASSUME INFINITE RENEWAL
SALVAGE IS LOST BENEFIT (COST)

REPAIR: $EUAC_1 = (9000 + 13000)(A/P, 8\%, 20) + 500 - (10000)(A/F, 8\%, 20)$
$EUAC_1 = (22000)(0.1019) + 500 - 10000(0.0219)$
$EUAC_1 = \$2522.80/\text{year}$

REPLACE: $EUAC_2 = (40000)(A/P, 8\%, 25) + 100 - 15000(A/F, 8\%, 25)$
$EUAC_2 = (40000)(.0937) + 100 - (15000)(0.0137)$
$EUAC_2 = \$3642.50$
repair bridge

[13] $D = 150000/15 = \$10000/\text{year}$
$P = -I + R(1-t) - C(1-t) + D(t)$
$PW = 0 = -150000 + (32000)(1-.48)(P/A, i\%, 15) - 7530(1-.48)(P/A, i\%, 15) + 10000(.48)(P/A, i\%, 15)$
$150000 = [16640 - 3915.6 + 4800](P/A, i\%, 15)$
$(P/A, i\%, 15) = 8.5595$
$i = 8\%$ from tables

[14]
(a) $\dfrac{1{,}500{,}000 - 300{,}000}{1{,}000{,}000} = 1.2$

(b) $1500000 - 300000 - 1000000 = \$200{,}000$

[15] ASSUME RENOVATION OCCURS AT $t=0$
$0 = P = (14000 + 1000) + (75(12) - 150 - 250)(P/A, 10\%, 10) + S(P/F, 10\%, 10)$
$0 = -15000 - 500(6.1446) + S(.3855)$
$11927.7 = (.3855)S$
$S = \$30940.86$

[16] $P = A(P/A, i\%, 30)$
$(P/A, i\%, 30) = 2000/89.30 = 22.396$
$i = 2\%$ per month
$i = (1+\phi)^K - 1$
$i = (1+.02)^{12} - 1 = .2682$ or 26.82%

[17]
$SL{:}\ D = \dfrac{I-S}{t} = \dfrac{500{,}000 - 100{,}000}{25} = \$16{,}000$

SOYD: $T = \frac{1}{2}(25)(26) = 325$
$D_1 = \frac{25}{325}(500{,}000 - 100{,}000) = \$30{,}769$
$D_2 = \frac{24}{325}(400000) = \$29{,}538$
$D_3 = \frac{23}{325}(400000) = \$28{,}308$

DDB: $D_1 = \frac{2}{25}(500{,}000) = \$40000$
$D_2 = \frac{2}{25}(500000 - 40000) = \$36{,}800$
$D_3 = \frac{2}{25}(500000 - 40000 - 36800) = \$33{,}856$

[18] PW = −12000 − 1000 (P/A, 10%, 10)
− 200 (P/G, 10%, 10)
+ 2000 (P/F, 10%, 10)

PW = −12000 − 1000 (6.1446)
− 200 (22.8913)
+ 2000 (.3855)

PW = −$21,951.86

EUAC = 21951.86 (A/P, 10%, 10)
= 21951.86 (0.1627) = $3571.56/year

[19] if the probability of failure for a life of N is 1/N each year

$EUAC_9 = 1500(A/P, 6\%, 20) + \left(\frac{1}{9}\right)(.35)1500 + (.04)1500$
$= 1500\left[.0872 + \left(\frac{1}{9}\right)(.35) + .04\right] = \$249.13$

$EUAC_{14} = 1600\left[.0872 + \left(\frac{1}{14}\right)(.35) + .04\right] = \$243.52$

$EUAC_{30} = 1750\left[.1272 + \frac{1}{30}(.35)\right] = \$243.01$

$EUAC_{52} = 1900\left[.1272 + \frac{1}{52}(.35)\right] = \$254.47$

$EUAC_{86} = 2100\left[.1272 + \frac{1}{86}(.35)\right] = \$275.64$

30 year pipe lowest cost

[20] $EUAC_7 = 0 \cdot (A/P, 10\%, 20) + .15(25000)$
$EUAC_7 = 0 \cdot (.1175) + 3750 = \$3750$
$EUAC_8 = 15000(.1175) + .10(25000) = \$4262.50$
$EUAC_9 = 20000(.1175) + .07(25000) = 4100.00$
$EUAC_{10} = 30000(.1175) + .03(25000) = \$4275.$

do nothing

[21] [a]
$EUAC_1 = 10000(A/P, 20\%, 1) + 2000 - 8000(A/F, 20\%, 1)$

$EUAC_1 = 10000(1.20) + 2000 - 8000(1.0) = \$6000$

$EUAC_2 = 10000(A/P, 20\%, 2) + 2000 + 1000(A/G, 20\%, 2) - 7000(A/F, 20\%, 2)$

$EUAC_2 = 10000(.6545) + 2000 + 1000(.4545) - 7000(.4545) = \$5818.$

$EUAC_3 = 10000(A/P, 20\%, 3) + 2000 + 1000(A/G, 20\%, 3) - 6000(A/F, 20\%, 3)$

$EUAC_3 = 10000(.4747) + 2000 + 1000(.8791) - 6000(.2747)$

$EUAC_3 = \$5977.90$

$EUAC_4 = 10000(A/P, 20\%, 4) + 2000 + 1000(A/G, 20\%, 4) - 5000(A/F, 20\%, 4)$

$EUAC_4 = 10000(.3863) + 2000 + 1000(1.2742) - 5000(.1863) = \$6205.70$

$EUAC_5 = 10000(A/P, 20\%, 5) + 2000 + 1000(A/G, 20\%, 5) - 4000(A/F, 20\%, 5)$

$EUAC_5 = 10000(.3344) + 2000 + 1000(1.6405) - 4000(.1344) = \$6446.40$

SELL AT END OF 2nd YEAR

[b] 6000 + (5000 − 4000) = $7000

[22] costs due to business

<u>maint</u>: 200 − 150 = $50/year
<u>insurance</u>: 300 − 200 = $100/year
<u>salvage</u>: (1000 − 500)(A/F, 10%, 5) = $81.90/year
<u>gasoline</u>: 5000(.6)/15 = $200/year

$\text{cost/mile} = \frac{100 + 50 + 81.9 + 200}{5000} = \$0.0864$

[a] yes $0.10/mile is adequate

[b] $[.1][X] = 5000(A/P, 10\%, 5) + 250 + 200 - 800(A/F, 10\%, 5) + \frac{X}{15}(.60)$

$.1X = 1637.96 + .04X$
$X = 27299$ miles/year

[1] no external force moves so $\Delta W = 0$ and $\Delta Q = 0$

then from first law:

$$\Delta U_1 + \Delta U_2 = 0$$

since $\Delta U = n C_v \Delta T$

$$n_1 C_v (T - T_1) + n_2 C_v (T - T_2) = 0$$

or

$$T = \frac{n_1 T_1 + n_2 T_2}{n_1 + n_2} \qquad (1)$$

ideal gas

$$n_1 = \frac{P_1 V_1}{R T_1} \qquad n_2 = \frac{P_2 V_2}{R T_2}$$

substitution into results in

$$T = T_1 T_2 \frac{P_1 V_1 + P_2 V_2}{[P_1 V_1 T_2 + P_2 V_2 T_1]}$$

$T_1 = 473°K$; $T_2 = 573°K$; $P_1 = 8$ atm; $P_2 = 6$ atm

$V_1 = 9\ ft^3$ $\quad V_2 = 1\ ft^3$

$$T = (473)(573) \frac{8(9) + 6(1)}{[8(9)573 + 6(1)473]}$$

$$T = 479.4°K$$

$$P = \frac{(n_1 + n_2) R T}{V_1 + V_2} = \frac{P_1 V_1 + P_2 V_2}{V_1 + V_2}$$

$$\therefore P = \frac{8(9) + 6(1)}{9 + 1} = 7.8 \text{ atm}$$

[2] $\Delta U = n c_v \Delta T = n c_v (T_2 - T_1)$

$$\Delta U = n c_v \left(\frac{P_2 V_2}{R n} - \frac{P_1 V_1}{R n} \right)$$

$$\Delta U = c_v \left(\frac{P_2 V_2 - P_1 V_1}{R} \right)$$

ideal gas $C_p - C_v = R$ $\therefore C_v \sim 5$

$$\Delta U = 5 \left(\frac{119(.01) - 115(.01)}{10.73} \right)$$

$$\Delta U = 0.0186 \text{ Btu}$$

[3] since for ideal gas

$$C_p - C_v = R$$

C_p is larger

[4] $\Delta S = \frac{\Delta Q}{T}$ phase change

$$\Delta S = \frac{96}{368.5} = 0.2605 \frac{\text{cal}}{°K \text{ gmole}}$$

$\Delta W = 0$ $\quad \Delta T = 0$ from first law:

$\therefore \Delta U = 0$

[5] for 1 gmole

$$\Delta H = (1) C_{p_2} T_2 - (1)\Lambda - (1) C_p T_1$$

$$\Delta H = (1)(9)(-2) - (1)(1436) - (1)(18)(-2)$$

$$\Delta H = -1418 \text{ cal}$$

$$\Delta S = \frac{\Delta Q}{T} = \frac{-1436 \text{ cal}}{271°K} = -5.299 \frac{\text{cal}}{°K}$$

[6]

$$-\frac{\Delta H}{R} = \frac{\ln K_2 - \ln K_1}{\frac{1}{T_2} - \frac{1}{T_1}}$$

$$\frac{-\Delta H}{R} = \frac{\ln(3992) - \ln(.0002)}{\frac{1}{1273} - \frac{1}{873}} = -46701\ °K$$

$$\Delta H = 46701(1.987) = 92796 \text{ cal/mole}$$

$$-46701 = \frac{\ln(3992) - \ln K_{900}}{\frac{1}{1273} - \frac{1}{1173}}$$

$$K_{900} = 175$$

$I_2 \rightarrow 2I$ $\quad$ let x = moles I_2 diss.
1 mole I_2 start

$$K = \frac{(2x)^2}{1 - x} \Rightarrow x^2 + \frac{K}{4}x - \frac{K}{4} = 0 = 4x^2 + Kx - K$$

$$x = \frac{-175 + \sqrt{175^2 + 16(175)}}{8} = .978$$

[7] Van der Waals

$$\left(P+\frac{a}{V^2}\right)(V-b)=RT$$

solve for PV/RT

$$\frac{PV}{RT}=Z=\frac{V}{V-b}-\frac{a}{RTV}$$

$$Z=\frac{1}{1-b/V}-\frac{a/RT}{V}$$

from binomial series from algebra

$$\frac{1}{1-x}=1+x+x^2+x^3+\cdots \quad (x^2<1)$$

$$\therefore Z=1+\frac{b}{V}-\frac{a/RT}{V}+\frac{b^2}{V^2}+\frac{b^3}{V^3}+\cdots$$

virial form

$$Z=1+\frac{b-a/RT}{V}+\frac{b^2}{V^2}+\frac{b^3}{V^3}+\frac{b^4}{V^4}+\cdots$$

[8] from appendix C,

for all T_R if $P_R>7.5$ $Z>1$

[9] in mathematics,

if $y=f(x,z)$

and $\frac{\partial y}{\partial x}=0$ then $y=f(z)$ only

if $H=f(T,P)$ and $U=f(T,V)$

and $dU=\left(\frac{\partial U}{\partial T}\right)_V dT+\left(\frac{\partial U}{\partial V}\right)_T dV$

$$dH=\left(\frac{\partial H}{\partial T}\right)_P dT+\left(\frac{\partial H}{\partial P}\right)_T dP$$

it can be shown from the definition of U that

$$dU=TdS-PdV$$

from Maxwell equations

$$\left(\frac{dU}{dV}\right)_T=T\left(\frac{\partial P}{\partial T}\right)_V-P$$

for ideal gas: $PV=RT$

$$\left(\frac{\partial P}{\partial T}\right)_V=\frac{P}{T}$$

$$\left(\frac{\partial U}{\partial V}\right)_T=T\left(\frac{P}{T}\right)-P=0 \quad \text{(ideal gas)}$$

$$\therefore dU=\left(\frac{\partial U}{\partial T}\right)_V dT$$

$U=f(T)$ only (ideal gas)

likewise

$$\left(\frac{\partial H}{\partial P}\right)_T=V-T\left(\frac{\partial V}{\partial T}\right)_P \quad \text{(any gas)}$$

$$\left(\frac{\partial V}{\partial T}\right)_P=\frac{V}{T} \quad \text{ideal gas}$$

$$\left(\frac{\partial H}{\partial P}\right)_T=V-T\left(\frac{V}{T}\right)=0$$

$H=f(T)$ only (ideal gas)

by definition

$$C_V=\left(\frac{dU}{dT}\right)_V;\quad C_P=\left(\frac{\partial H}{\partial T}\right)_P$$

can be shown that

$$\left(\frac{\partial C_V}{\partial V}\right)_T=T\left(\frac{\partial^2 P}{\partial T^2}\right)_V \quad \text{any gas}$$

$$\left(\frac{\partial C_P}{\partial P}\right)_T=-T\left(\frac{\partial^2 V}{\partial T^2}\right)_P \quad \text{any gas}$$

for ideal gas

$$\left(\frac{\partial^2 P}{\partial T^2}\right)_V=0 \quad \left(\frac{\partial^2 V}{\partial T^2}\right)_P=0 \quad \text{ideal gas}$$

$$\therefore \left(\frac{\partial C_V}{\partial V}\right)_T=\left(\frac{\partial C_P}{\partial P}\right)_T=0 \quad \text{ideal gas}$$

$\therefore C_P=f(T)$ only ideal gas

$C_V=f(T)$ only ideal gas

[10] since at infinite dilution: $x_1=x_2=0$

$A_{12}=\ln(2)=0.6932$

$A_{21}=\ln(.5)=-0.6932$

[11] PROCESS:

COOL GAS: 500°F → −33.4°C (1)
CONDENSE: −33.4°C (2)
COOL LIQ: −33.4°C → −77.7°C (3)
SOLIDIFY: −77.7°C (4)
COOL LIQ: −77.7°C → −150°C

$T_1 = 500°F = 533.1°K$

$T_2 = -33.4°C = 239.8°K$

$T_3 = -77.7°C = 195°K \qquad T_4 = -150°C = 172°K$

STEP (1) eq. 4.59 for 1 gmole:

$$\Delta S_{gas} = 1\left[6.5486 \ln\frac{239.8}{533.1} + 6.1251\times10^{-3} \times(239.8-533.1) + 2.3663\times10^{-6}\frac{(239.8^2-533.1^2)}{2} + -1.5981\times10^{-9}\frac{(239.8^3-533.1^3)}{3}\right]$$

$\Delta S_{gas} = -7.2568$ cal/gmole°K

$= \frac{-7.2568}{17.032} = -0.426$ cal/gram°K

step(2) eq 4.55

$\Delta S_c = \frac{-5581}{239.8}(1) = -23.274$ cal/gmole°K

$= -1.3665$ cal/gram°K

step(3) eq 4.58

$\Delta S_L = 1.06 \ln\frac{195.5}{239.8} = -0.2165$ cal/gram°K

step(4) eq 4.55

$\Delta S_f = \frac{-1352}{195.5} = -6.9156$ cal/gmole°K

$= -0.406$ cal/gram°K

cool solid: eq 4.58

$\Delta S_s = 0.502 \ln\frac{172}{195.5} = -0.0643$ cal/g°K

$\Sigma\Delta S = \Delta S_g + \Delta S_c + \Delta S_L + \Delta S_f + \Delta S_s$

$= -0.426 - 1.3665 - 0.2165 - 0.406 - 0.064$

$= -2.479$ cal/g°K $= -2.479$ Btu/lb°R

[b] cool gas 25°C → −33.4°C

$T_1 = 298.2°K \quad T_2 = 239.8°K$

$$\Delta S_{gas} = 1\left[6.5486 \ln\frac{239.8}{298.2} + 6.1251\times10^{-3} \times\frac{(239.8^2-298.2^2)}{2} - 1.5981\times10^{-9}\frac{(239.8^3-298.2^3)}{3}\right]$$

$\Delta S_{gas} = -1.8366$ cal/gmole°K

$= -0.1078$ cal/g°K

from part [a]

$\Delta S = -0.1078 - 1.3665 - 0.2165 - 0.4060$

$= -2.09868$ cal/g°K

$S_{298} = 46.03$ cal/gmole $= 2.7026$ cal/g°K

since $\Delta S = S_{195.5} - S_{298}$

$S_{195.5} = -2.09868 + 2.7026$

$= 0.6058$ Btu/lb°R

[12]

ASSUME: MECH ENERGY → HEAT
NO HEAT LOSS
PUMP & MIXER 100% EFF.

MW $Na_2C_2H_3O_2 = 82$

1 lbmole:		wgt %
H_2O:	.91×18 = 16.4 lbs	16.4/23.8 = 69%
$Na_2C_2H_3O_2$:	.09×82 = 7.4 lbs	7.4/23.8 = 31%
	23.8 lb	

solution pumped:

20 gpm × 60 min/hr × 8.33 lb/gal × 1.15 = 11500 lb/hr

rates: $Na_2C_2H_3O_2$: .31 (11500) = 3565 lb/hr
H_2O: .69 (11500) = 7935 lb/hr

heat of solution:

$-3943 \frac{cal}{mole} \times \frac{1\ mole}{82\ g} \times \frac{3.960\times10^{-3}\ Btu}{cal} \times 453 \frac{gram}{lb}$

$= -86.5$ Btu/lb

Using 55°F as datum:

Heat balance

$(11500)(0.94)(135-55) = 3565(0.339)(68-55) + 3565(86.5) + (5+7.5)(2544 \frac{Btu}{hr\cdot hp}) + Q$

Q = 509,000 Btu/hr added

[13] process is adiabatic compression of a gas

P_1, T_1 : initial gas temperature

$$P_2 = P_1 (T_2/T_1)^{\gamma/(\gamma-1)}$$

$\gamma = 1.4$ (air)

$T_1 = 294°K$, $T_2 = 573°K$

$P_1 = 1$ atm

$$P_2 = (1)[573/294]^{\frac{1.4}{.4}} = 10.4 \text{ atm}$$

[14] weight flow:

$$1.2 \text{ gpm} \times \frac{1 \text{ ft}^3}{7.5 \text{ gal}} \times 62.4 \frac{\text{lb}}{\text{ft}^3} = 9.98 \frac{\text{lb}}{\text{min}}$$

[a] $h_{steam} = 1150.4$ Btu/lb (tables)

$h_f = 18.07$ Btu/lb 50°F

$h_f = 180.07$ Btu/lb 212°F

$h_f = 117.89$ Btu/lb 150°F

Δh balance let x = lbs steam

$$9.98(18.07) + x(1150.4) = (9.98 + x)117.89$$

$x = 0.964$ lb/min steam

[b]

$$Q = 9.98(1)(117.89 - 18.07) = 996.2 \text{ Btu/min}$$

$$W_{STEAM} = \frac{996.2}{(1150.4 - 180.07)} = 1.03 \text{ lb/min}$$

[15]

		$\Delta H°$ cal
$CO + \frac{1}{2}O_2 \rightarrow CO_2$		-67636
$H_2 + \frac{1}{2}O_2 \rightarrow H_2O_l$		-68313
$H_2O_l \rightarrow H_2O_g$		$+10519$
$H_2 + CO + O_2 \rightarrow CO_2 + H_2O$		-125430

moles $N_2/O_2 = 79/21 = 3.76$

total moles before reaction:

$$1_{O_2} + 1_{CO} + 1_{H_2} + 3.76_{N_2} = 6.76$$

after:

$$1_{CO_2} + 1_{H_2O} + 3.76_{N_2} = 5.76$$

$\Delta H = \Delta U + \Delta PV$

$\Delta U = 0$: $\Delta H = \Delta PV$

ideal gas

$$\Delta PV = n_2 R T_2 - n_2 R T_1 = R(n_2 T_2 - n_1 T_1)$$

$$\Delta H = \Delta H_{298} + (\bar{C}_{p_{CO_2}} + C_{p_{H_2O}} + 3.76 \bar{C}_{p_{N_2}})(T_2 - T_1) \quad (a)$$

$$\Delta H = R[5.76 T_2 - 6.76 T_1] \quad (b)$$

i	C_{p_i}	$n_i C_{p_i}$	
CO_2	13.5	13.5	assume 3000°K
H_2O	11.0	11.0	
N_2	8.25	31.02	
		55.52	

above

if $T_2 = 3000°K = 5400°R = 4940°F$

combining (a) & (b)

$$-125430 + 55.52(T_2 - 298) = 1.987 \times [5.76 T_2 - 6.76(298)]$$

$T_2 = 3131°K = 2858°C$

close enough. otherwise recalc. table at 3100

$$P_2 = \frac{n_2 T_2}{n_1 T_1} P_1 = \frac{5.76}{6.76} \frac{3131}{298} \times 5$$

$P_2 = 44.8$ atm

[16] air lift work:

$$W_a = 20\,\text{gpm}(8.33\times1.5)\frac{\text{lb}}{\text{gal}}\times 50\,\text{ft}$$
$$= 12500\ \text{ft-lb/min}$$

isothermal expansion:

$$W_a = nRT_1 \ln\frac{P_1}{P_2} = P_1V_1\ln\frac{P_1}{P_2}$$

actual pump (air lift) work

$$W_a = 12500/.3 = 41700\ \text{ft-lb/min}$$

$$\therefore\ 41700 = (14.7\times144)\frac{\text{lb}}{\text{ft}^2}\left[\ln\frac{64.7}{14.7}\right]V_1$$

$$V_1 = 13.27\ \text{ft}^3/\text{min}$$

isentropic work for compressor:

$$-W_c = \frac{nRT_1}{\gamma-1}\gamma\left[\left(\frac{P_2}{P_1}\right)^{\gamma-1/\gamma} - 1\right]$$

$$-W_c = \frac{P_1V_1\gamma}{\gamma-1}\left[\left(\frac{P_2}{P_1}\right)^{\gamma-1/\gamma} - 1\right]$$

$$-W_c = \frac{(14.7)(144)(13.27)(1.4)}{.4}\left[\left(\frac{64.7}{14.7}\right)^{.4/1.4} - 1\right]$$

$$-W_c = 52000\ \text{ft-lb/min}$$

$$\therefore\ \text{pump hp: } \frac{52000}{3300} = 1.57\ \text{hp}$$

[17]

assume $K_a = K_p$: $FeO + CO \rightarrow FeS + CO_2$

$$K_p = \frac{y_{CO_2}}{y_{CO}}P^{1-1} = \frac{y_{CO_2}}{y_{CO}} = 40$$

solved two ways:

BASIS: 1 mole gas enters	pure CO
$CO = 0.2 - X$	$CO = 1 - X$
$CO_2 = X$	$CO_2 = X$
$40 = \frac{X}{0.2-X}$	$40 = \frac{X}{1-X}$
$X = 0.195$	$X = 0.976$
moles Fe = 0.195	moles Fe = 0.976

[18]

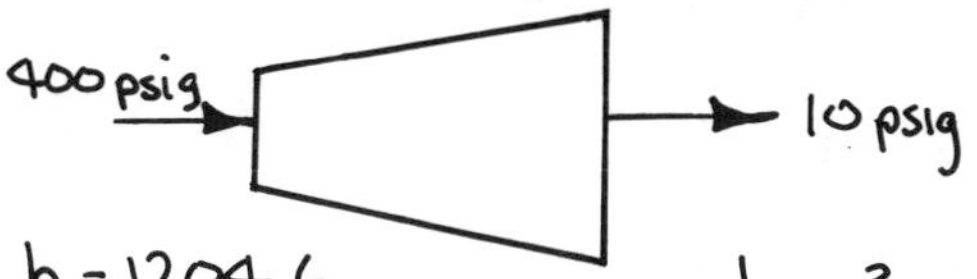

$h_1 = 1204.6$ $\quad$ $h_2 = ?$

$S_1 = 1.481$ $\quad \Delta S = 0 \quad$ $S_2 = 1.481$

$T_1 = 448°F$ $\quad$ $S_f = 0.3523$

x = liquid fraction $\quad$ $S_v = 1.7149$

$$1.481 = x(.3523) + (1-x)(1.7149)$$

$X = .172$ or 17.2% liquid

[a] for every 100 lb steam 82.8 lb steam is available

[b] 100%, on the basis that the heat of condensation is deducted and result is net work

[c] if $T_1 = 548°F$ $\quad S_1 = 1.5541$

$$1.5541 = x(.3523) + (1-x)(1.7149)$$

$X = .118$ or 11.8%

thus 88.2 lbs process steam available per 100 lbs higher pressure steam.

[19] acetonitrile

$$T_R = \frac{550+460}{548(1.8)} = 1.02$$
$$P_R = \frac{4500}{14.7(47.7)} = 6.4$$

} $Z = 0.801$ from generalized plots

nitrogen

$$T_R = \frac{1010}{126.2(1.8)} = 4.45$$
$$P_R = \frac{10}{33.5} = .3$$

} $Z = 1.0$

$$m_a = \frac{PV}{ZRT} = \frac{4500(0.2)}{0.801(10.73)(1010)} = 0.1037\ \text{lbmoles}$$

$$m_N = \frac{10(2)}{(1)(.73)(1010)} = .0271\ \text{lbmoles}$$

$$y_a = .1037/(.1037 + .0271) = 0.793$$

[19] con't

using pseudocritical rule of Kay

$T_{PC} = 0.793(548) + 0.207(126.2) = 461\,^\circ K$

$P_{PC} = 0.793(47.7) + 0.207(33.5) = 44.8$ atm

$T_R = \dfrac{1010}{461(1.8)} = 1.217$

$\dfrac{P}{P_{PC}} = \dfrac{ZnRT}{P_{PC}V} = P_R = \dfrac{Z\,.1308(.73)\,1010}{44.8\,(2.2)}$

$P_R = Z\,(0.992)$ at $T_R = 1.217$

using generalized charts: at $T_R = 1.217$

assume Z	P_R	Z (from table)
0.9	.893	.817
0.85	.843	.83
0.84	.846	.84

$\therefore P_r = 0.846$

$P = 0.846(44.8) = 37.9$ atm

[20] $h_A = 60$

$h_E = 1354$, $h_F = 1192$,

$h_G = 1378$, $h_H = 984$

$h_{F'} = 878$

[a] $Q_1 = (h_E - h_A) + (h_G - h_F) = (1354 - 60) + (1378 - 1192) = 1480$ Btu/lb

[b] $Q_2 = h_A - h_H = 60 - 984 = -924$ Btu/lb

[c] $W_{NET} = Q_1 + Q_2 = 556$ Btu/lb

[d] $\eta = W_{NET}/Q_1 = 556/1480 = 0.376$

[e) quality = 100 − 11.2 = 88.8% (Mollier Chart)

RANKINE:

[a] $Q_1 = h_E - h_A = 1294$ Btu/lb

[b] $Q_2 = h_A - h_{F'} = -818$ Btu/lb

[c] $W_{NET} = Q_1 + Q_2 = 476$ Btu/lb

[d] $\eta = W_{NET}/Q_1 = 0.368$

[e] quality = 100 − 21.4 = 78.6%

[21] at azeotrope

$\gamma_i = \dfrac{P}{P_i^\circ}$ or $\ln\gamma_i + \ln P_i^\circ = \ln P$

if we let $e = \dfrac{A_{12}}{A_{21}}\dfrac{X_1}{1 - X_1}$ then

$\ln\gamma_1 = \dfrac{A_{12}}{(1+e)^2}$ $\quad \ln\gamma_2 = \dfrac{A_{21}}{(1+\frac{1}{e})^2}$

$\dfrac{A_{12}}{(1+e)^2} = \ln P - \ln P_1^\circ$

$\dfrac{A_{21}}{(1+\frac{1}{e})^2} = \ln P - \ln P_2^\circ$

$\therefore \dfrac{A_{21}}{A_{12}} e^2 = K = \dfrac{\ln P - \ln P_2^\circ}{\ln P - \ln P_1^\circ}$

$\therefore X_1 = \dfrac{\sqrt{\frac{A_{21}}{A_{12}}K}}{1 + \sqrt{\frac{A_{21}}{A_{12}}K}}$

Calculation scheme:

guess t, calculate P_1° & P_2°, if P_1° & P_2° are above 760 mm calculate X_1, check that $\Sigma y_i = 1$ from $\dfrac{\gamma_1 X_1 P_1^\circ + \gamma_2 X_2 P_2^\circ}{P} = 1$

t	P_1°	P_2°	K	X_1	e	γ_1	γ_2	Σy_i
50	612	507						
55	730	601						
60	863	709						
65	1017	832	.311	.329	.634	.848	.950	1.07
64	985	806	.226	.295	.542	.831	.958	1.03
63	953	781	.227	.295	.542	.831	.958	1.062

$t = 63\,^\circ C \quad X_{ACETONE} = 0.295$

[1]

$V_{BOX} = 1\text{ ft}^3;\ V_{H_2O} = 1\text{ ft}^3$

$W_{BOX} = \rho_{IRON} V_{IRON};\ \rho_{IRON} = 62.45(7.2) = 449.7$

$W_{H_2O} = \rho_{H_2O} V_{H_2O} = 62.45(1) = 62.45\text{ lb}$

$V_{IRON} = 1 - \left(\frac{11.5}{12}\right)^3 = 0.1198\text{ ft}^3$

$W_{BOX} = (449.7)(0.1198) = 53.895\text{ lb}$

floats because equal volume of box weighs less than water.

$\%\text{ submergence} = \frac{53.895}{62.45} \times 100 = 86\%$

[2] $h_L = f\left(\frac{L}{D}\right)\frac{v^2}{2g}$ f is unchanged since fully turbulent

$$\frac{h_1}{h_2} = \frac{f_1 \frac{L_1}{D_1} \frac{v_1^2}{2g}}{f_2 \frac{L_1}{D_1} \frac{v_2^2}{2g}}$$

$\therefore \frac{v_1^2}{v_2^2} = \frac{1}{2}$ $v_2 = 1.4 v_1$

flow increased by 40%

[3] $v = \sqrt{2g\Delta h}$ $h_{air} = \frac{\frac{1}{12}\rho_{Hg}}{\rho_{air}} = \frac{\frac{1}{12}(13.56)62.4}{0.0764} = 925.7\text{ ft air}$

$v = \sqrt{2(32.2)925.7} = 244.2\text{ fps air}$

$h_{H_2O} = \frac{\frac{1}{12}(13.56)62.4}{61.82} = 1.143\text{ ft H}_2\text{O}$ $\rho_{H_2O} = 61.862$ @ 110°F

$v = \sqrt{2(32.2)(1.143)} = 8.58\ \frac{\text{ft}}{\text{sec}}\ H_2O$

[4] $h_{pump} = h_{Loss} + h_s + h_v + h_{outlet}$

$h_s = 100\text{ ft}$

$h_v = 0$

$h_{Loss} = .0311 f \frac{LQ^2}{d^5}$

$v = 0.408 \frac{Q}{d^2} = 0.408 \frac{750}{6.065^2} = 8.319\ \frac{\text{ft}}{\text{sec}}$

$$N_{Re} = \frac{Dv\rho}{\mu_e} = \frac{\frac{6.065}{12}(8.319)(0.9 \times 62.45)}{0.0015}$$

$N_{Re} = 1.57 \times 10^5 \Rightarrow f = 0.0185$ {Crane A-24}

$$h_{Loss} = .0311 f \frac{L}{d^5} Q^2 = (0.0185)(0.0311) \times \frac{3000}{6.065^5}(750)^2$$

$h_{Loss} = 118\text{ ft};\ h_{outlet} = \frac{50(144)}{.9(62.45)} = 128$

$h_{pump} = 118 + 100 + 0 + 128 = 346\text{ ft oil}$

$p = \frac{346(.9 \times 62.45)}{144} = 135\text{ psi}$

[5] $\frac{L}{D} = R_T + (n-1)\left(R_L + \frac{R_B}{2}\right)$

{Crane manual}

$r/d = 3/.21 = 14.3$

$\therefore \frac{L}{D} = 39 + (47)\left(22 + \frac{17}{2}\right) = 1472.5\text{ diam.}$

[6] $\frac{h_1}{h_2} = \frac{64/N_{Re1}}{64/N_{Re2}} = \frac{N_{Re2}}{N_{Re1}} = \frac{\mu_1}{\mu_2}$

$\frac{h_1}{h_2} = \frac{1}{0.5} = 2$

$h_2 = 0.5 h_1$ pressure drop decreases $\frac{1}{2}$

[7] @ 20 gpm in 1-inch sch 40 pipe:

Crane: B-14 $\Delta P = 10.9\text{ psi}/100\text{ ft}$

$\Delta P = 10.9\left(\frac{65}{100}\right) = 7.09\text{ psi}$

[8] Crane: B-15 @ $100\ \frac{\text{ft}^3}{\text{min}}$: $\Delta P = 0.534\ \frac{\text{psi}}{100\text{ ft}}$

ideal gas correction

$$\Delta P = \overset{0.8\times}{0.534}\left[\frac{100 + 14.7}{112 + 14.7}\right]\left[\frac{460 + 95}{520}\right] = 0.418\text{ psi}$$

[9] EQUIV LENGTHS SCREWED IN FEET (TABLE 5.11)

ELLS: $3 \times 13 = 39$ ft

COUPLINGS: $4(.65) = 2.6$ ft

[Crane A30]

GATE VALVE ½ OPEN

$L/D = 160 \quad L = 160(4.026/12) = 53.7$ ft

Perry 5th Ed. Table 5-19

PLUG COCK $\theta = 20°$: $L/D = 1.56(45)$

$L = 1.56(45)(4.026/12) = 23.6$

$L_{TOTAL} = 400 + 200 + 39 + 2.6 + 53.7 + 23.6$

$L = 718.9$ ft

Crane B-14

$\Delta P = 5.65$ psi/100 ft

$\Delta P = 5.65\left(\frac{718.9}{100}\right) = 40.6$ psi

[10]

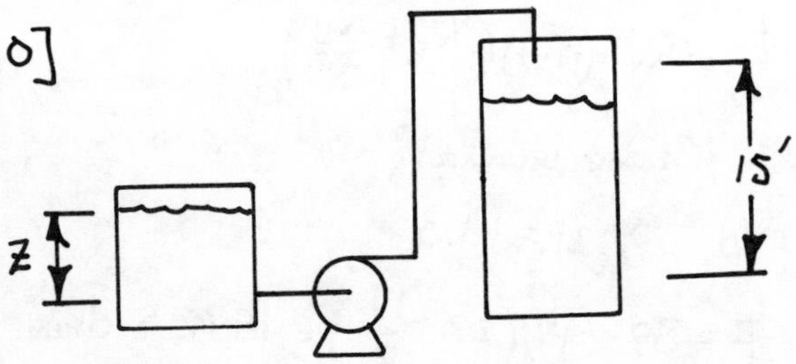

z, 15′

CRANE:

sched 40 2″ PIPE: 1 ft/sec = 10.45 gpm

ID = 2.067 in

A = 0.0233 ft²

Bernoulli's: $z + h_{pump} = 15 + \frac{V^2}{2g} + h_f$

$h_{pump} = \left(\frac{1}{8}\text{ hp}\right)\left(550\frac{\text{ft-lb}}{\text{hp-sec}}\right)\left(\frac{1}{5}\frac{\text{min}}{\text{ft}^3}\right)\left(\frac{1}{62.4}\frac{\text{ft}^3}{\text{lb}}\right)\left(60\frac{\text{sec}}{\text{min}}\right)$

$= 13.22$ ft

$h_f = 0.8$ ft $\quad V = 5\frac{\text{ft}^3}{\text{min}}\, 7.48\frac{\text{gal}}{\text{ft}^3}\, \frac{1\text{ ft/sec}}{10\text{ gal/min}}$

$V = 3.57$ ft/sec

$z = -13.22 + 15 + \frac{3.57^2}{2(32.2)} + 0.8$

$z = 2.8$ ft

[11] Crane Tech Paper #410 9th PRINTING page A-27

Relative radius $= \frac{72/2}{3} = 12$

EQUIV. LENGTH OF 1-90° BEND = 34.5 diam.

R_t = total resistance 1 - 90° bend

n = total number 90° bends

R_L = resistance due to length

R_B = resistance due to bend

$$\frac{L}{D} = R_t + (n-1)\left(R_L + \frac{R_B}{2}\right)$$

$R_L = 18.7$

$R_B = 15.2$

$n = 60$

$\frac{L}{D} = 34.5 + (60-1)\left[18.7 + \frac{15.2}{2}\right] = 1600$ diam

$L = 1600\left[\frac{3 - 2(.065)}{12}\right] = 384$ ft

$N_{Re} = 50.6\frac{Q\rho}{d\mu}$ $\quad Q = 50$ gpm, $d = 2.87''$, $\rho = 62.4$ lb/ft³, $\mu = 0.862$ cp

$N_{Re} = 50.6\frac{(150)(62.4)}{(2.87)(0.862)} = 190000$

$f = 0.0195$ Moody chart

$\Delta P = 0.000216\, f L \rho Q^2/d^5$

$\Delta P = 0.000216(.0195)\frac{(384)(62.4)(150^2)}{2.87^5}$

$\Delta P = 11.7$ psi

[12] $N_{Re} = 123.9\, dv\rho/\mu$

$= 123.9(1)\frac{(7.33 \times 55)}{25} = 2000$ laminar

$f = \frac{64}{N_{Re}} = .032$

New product: $N_{Re} = 2000\frac{25}{12.5} = 4000$

$f_{new} = .0392$ (Moody chart)

increase in h.p. proportional to f:

% increase h.p $= \frac{0.0392 - .032}{.032} \times 100$

$= 22.5\%$

[13] $W = 450{,}000 + 50{,}000 = 500{,}000$ lb/hr

Solution: $100 \times \frac{450{,}000}{500{,}000} = 90\%$

90% solution sp.gr. = 1.2   10% crystals sp gr 2.1

$$\text{sp.gr. slurry} = \frac{1}{\frac{1}{1.2} + \frac{1}{2.1}} = 1.254$$

$$C_v = 400 = Q\sqrt{\text{sp.gr.}/\Delta P}$$

Bernoulli

$$Z_1 + \frac{P_1}{\rho} = h_{pump} + h_f + h_{valve} + Z_2 + \frac{P_2}{\rho}$$

$$Q = 5\times10^5 \frac{\text{lb}}{\text{hr}} \times \frac{1}{62.4(1.254)}\frac{\text{ft}^3}{\text{lb}} \times \frac{1}{60}\frac{\text{hr}}{\text{min}} \times 7.48 \frac{\text{gal}}{\text{ft}^3}$$

$= 797$ gal/min

$$C_v = 400 = 797\sqrt{\frac{1.254}{\Delta P}}$$

$\Delta P = 4.98$ psi across valve

$$\frac{\Delta P}{\rho} = h_{valve} = 4.98\times144/(62.4\times1.254)$$

$= 9.15$ ft

$$\frac{P_1 - P_2}{\rho} = h_A = -26'' \text{ Hg} = -26\left(\frac{13.5}{1.253}\right)\frac{1}{12}$$

$= -23.3$ ft

$$h_A = h_{valve} + h_{pump} + h_{friction} + h_{elevation}$$

$-h_{pump} = 23.3 + 9.15 + 18 + 30 = 80.45$ ft slurry

$-h_{pump} = 80.45(1.254) = 100.9$ ft H_2O

sign is negative for pump work on fluid

[14]

Bernoulli: $Z_1 + \frac{V_1^2}{2g} + \frac{P_1}{\rho} = Z_2 + \frac{v_2^2}{2g} + \frac{P_2}{\rho} + h_f$ (with $\frac{V_1^2}{2g} \to 0$, $Z_2 \to 0$)

$$Z_1 - \frac{v_2^2}{2g} - h_f = 0 \quad \text{since } P_1 - P_2 = 0$$

$$h_f = f\left(\frac{L}{D}\right)\frac{v^2}{2g}$$

$$\therefore \quad Z_1 - \frac{v_2^2}{2g}\left[1 + f\left(\frac{L}{D}\right)\right] = 0$$

$$v_2 = \sqrt{2gZ_1/(1 + f\tfrac{L}{D})}$$

$Q = v_2 A \qquad A = 0.0233 \text{ ft}^2$

$$Q = v_2(.0233)\left(7.48\frac{\text{gal}}{\text{ft}^3}\right)\left(60\frac{\text{sec}}{\text{min}}\right)$$

$Q = 10.46\, v_2$

$$\therefore \quad Q = 10.46\sqrt{2gZ_1/(1 + f\tfrac{L}{D})} \text{ gpm}$$

[15] $N_{Re} = 123.9\,\frac{dv\rho}{\mu}$

$$= 123.9\,\frac{(4.026)(10)(62.4)}{4.35}$$

$= 716{,}000$; $f = .0188$

$$h_f = f\left(\frac{L}{D}\right)\frac{v^2}{2g} = .0188\left(\frac{5280\times5\times12}{4.026}\right)\frac{10^2}{(2)(32.2)}$$

$h_f = 2297$ ft

$$\text{heat} = 2297 \text{ ft} \;\; 1.285\times10^{-3}\frac{\text{Btu}}{\text{ft-lb}}$$

$= 2.95$ Btu/lb

heat $= c_p \Delta T$

$$\Delta T = \text{heat}/c_p = \frac{2.95 \text{ Btu/lb}}{0.2 \text{ Btu/lb}^\circ\text{F}}$$

$\Delta T = 14.8^\circ$F temp rise

[16] Pressure drop due to friction loss

$$h_f = f\left(\frac{L}{D}\right)\frac{v^2}{2g}$$

Gas properties are a function of temp.

assume $\Delta P = 16$ psi

$$P_{Average} = \frac{60 + (60+16)}{2} = 68 \text{ psig}$$

$= 82.7$ psia

[16] con't

@ 82.7 psia, 70°F

$\rho_{CH_4} = 0.22 \frac{lb}{ft^3}$ $\mu = 7.39 \times 10^{-6} \frac{lb}{ft\text{-}sec}$

$v = \frac{Q}{A}$; $Q = 250\left(\frac{14.7}{82.7}\right) = 44.4 \frac{ft^3}{sec}$

$A = \pi 1^2/4 = 0.786 \text{ ft}^2$

$v = 44.4/0.786 = 56.58 \text{ ft/sec}$

$N_{Re} = \frac{Dv\rho}{\mu} = \frac{(1)(56.58)(.22)}{7.39 \times 10^{-6}} = 1.684 \times 10^6$

colbrook:

$f = \left[1.8 \log(N_{Re}/7)\right]^{-2} = 0.0106$

smooth pipe

$\Delta P = .001294 \frac{fL\rho v^2}{d} = .001294 (.0106)$

$\times \frac{3(5280)(.22)(56.58)^2}{12}$

$\Delta P = 12.8$

if $\varepsilon/D = .00010$

Shacham:

$f = -2 \log\left[\frac{\varepsilon/d}{3.7} - \frac{5.02}{N_{Re}} \log\left(\frac{\varepsilon/d}{3.7} + \frac{14.5}{N_{Re}}\right)\right]^{-2}$

$f = .0129$

$\Delta P = .001294 \frac{.0129(3 \times 5280)(.22)(56.58)^2}{12}$

$\Delta P = 15.6$ close enough

Reservior pressure $= 60 + 15.6 = 75.6$ psig

[17] Net pump head = Σelevation + Σpress. + Σveloc.

Σelevation = 4 ft

Σpress. head $= \left[30.7 - \left(14.7 + \frac{5}{29.92}(14.7)\right)\right] 2.3039$

Σpress. head = 31.2 ft

Σvel. $= (V_2^2 - V_1^2)/2g$

$V_2 = 818 \text{ gpm } (0.408)/5^2 = 13.3 \text{ ft/sec}$

$V_1 = V_2\left(\frac{5}{10}\right)^2 = 3.33 \text{ ft/sec}$

Σvelocity $= \frac{(13.3^2 - 3.33^2)}{2g} = 2.57$ ft

Net pump head $= 4 + 31.2 + 2.57 = 37.8$ ft

hyd hp $= \frac{(\text{gpm})(\text{head})(8.33)}{33000}$

$= \frac{(818)(37.7)(8.33)}{33000} = 7.78$ hp

eff% $= 100 \frac{\text{hyd hp}}{\text{input hp}} = 100 \frac{7.8}{10} = 78\%$

pump speed increased to 3500:

$Q_2 = Q_1\left(\frac{N_2}{N_1}\right) = 818 \frac{3500}{1750} = 1636$ gpm

$h_1 = h_2\left(\frac{N_2}{N_1}\right)^2 = 37.8\left(\frac{3500}{1750}\right)^2 = 151.2$ ft

$W_2 = W_1\left(\frac{N_2}{N_1}\right)^3 = 10\left(\frac{3500}{1750}\right)^3 = 80$ hp

[18] the pressure drop of compressible fluids is

$\Delta P = 3.36 \times 10^{-6} \frac{fW^2L}{d^5 \rho}$

EQUIV. LENGTH $= 23 + 30\,(6/12) = 68$ ft

(L/D for elbows)

$\Delta P = 100$ psi ; $d^5 = 6346 \text{ in}^5$; $\mu = 0.021$; $d = 5.761$

$\rho = (MW)P/RT$; $\rho = (28)(114.7)/[10.72(212+460)]$

$\rho = 0.4458 \text{ lb/ft}^3$

rearranging

$W = \sqrt{\Delta P d^5 \rho/(3.36 \times 10^{-6} fL)}$

$W = \sqrt{100(6346)(.4458)/(3.36 \times 10^{-6} f\, 68)}$

$W = 35188\sqrt{\frac{1}{f}}$

$N_{Re} = 6.31 W/(d\mu) = 6.31 W/(5.761 \times .021)$

$N_{Re} = 52.15 W$

$f = \left\{-2 \log\left[\frac{\varepsilon/D}{3.7} - \frac{5.02}{N_{Re}} \log\left(\frac{\varepsilon/D}{3.7} + \frac{14.5}{N_{Re}}\right)\right]\right\}^{-2}$

[18] con't

for steel pipe use $\varepsilon = .002$; ∴

$$\varepsilon/D = \frac{.002}{5.761/12} = 0.004$$

looking at fig 5.8

if $N_{Re} > 2.5\times10^5$ for $\varepsilon/D = .004$

f is constant.

trial & error:

1. guess f
2. calc. W
3. calc. N_{Re}
4. calc. f
5. compare: $f_{guess} : f_{calc.}$

f_{guess}	W	N_{Re}	$f_{calc.}$
0.0288	2.073×10^5	1.081×10^7	0.284
0.0284	2.087×10^5	1.089×10^7	0.284

∴ max CO_2 rate = 208700 lb/hr

[19] Bernoulli

$$Z_1 + \frac{P_1}{\rho} + \frac{V_1^2}{2g} = Z_2 + \frac{P_2}{\rho} + \frac{V_2^2}{2g} + h_f + h_{pump}$$

$Z_1 = 0$; $P_1 = P_2$; $V_1 = 0$, $Z_2 = 4000$ ft; $V_2 = 10$ ft/sec

$$h_{pump} = -\left(Z_2 + \frac{V_2^2}{2g} + h_f\right)$$

$$\frac{V_2^2}{2g} = \frac{100}{2(32.2)} = 1.6\text{ ft}$$

$$Q = vd^2/0.408 = \frac{10(5.761)^2}{.408} = 813\text{ gpm}$$

$\mu = 1.129$ cp

$$N_{Re} = 123.9\frac{dv\rho}{\mu} = 123.9(5.761)\frac{(10)(62.4)}{1.129}$$

$N_{Re} = 394811$

assume $\varepsilon/D = .0003$

using Shacham

$$f = \left\{-2\log\left[\frac{\varepsilon/d}{3.7} - \frac{5.02}{N_{Re}}\log\left(\frac{\varepsilon/d}{3.7} + \frac{14.5}{N_{Re}}\right)\right]\right\}^{-2}$$

$f = 0.0163$

$$h_f = f\left(\frac{L}{D}\right)\frac{v^2}{2g} = .0163\left(\frac{5000}{5.761/12}\right)1.6$$

$h_f = 271$ ft

$h_{pump} = -(4000 + 1.6 + 271) = -4273$ ft

$$\text{pump hp} = \frac{Q\,h_{pump}\,8.33}{33000}$$

$$\text{pump hp} = \frac{(813)4273(8.33)}{33000} = 877\text{ hp}$$

$Bhp = 876/.7 = 1252$ hp

$$\text{cost} = 1252\left(.745\frac{\text{KW}}{\text{hp}}\right)\left(\frac{\$.015}{\text{KW-hr}}\right)$$

$= \$14.00/\text{hr}$

[20]

Spherical particles

$\psi = 1$

∴ $\varepsilon = 1.08 - 1.12(1) + 0.405(1)^2$

$\varepsilon = 0.365$

$$V_{BED} = \left(\frac{\pi D^2}{4}\right)(L) = \frac{\pi 4^2}{4}(6) = 75.4\text{ ft}^3$$

$A = 12.6\text{ ft}^2$

flow = 20(75.4) = 1508 gpm

$$v_s = 1508(.1337)\left(\frac{1}{60}\right)\left(\frac{1}{12.6}\right)$$

$v_s = .2667$ ft/sec

$\rho = 62.4$ lb/ft^3

$\mu = (1.)(6.72\times10^{-4}) = 6.72\times10^{-4}$ lb/ft-sec

$$N^*_{Re} = \frac{(62.4)(.065/12)(.2667)}{6.72\times10^{-4}}$$

$$N^*_{Re} = 134.1$$

Ergun equation:

$$\Delta P = \frac{(1-.365)(62.4)(.2667)^2(6)}{(.365)^3\left(\frac{.065}{12}\right)(32.2)}\left[150\frac{(1-.365)}{134.1} + 1.75\right]$$

$$\Delta P = 1993.8\,[2.46] = 4905\ \text{lb/ft}^2$$

$$\Delta P = 34\ \text{psi}$$

[1]

$$\frac{q}{A} = \frac{t_1 - t_s}{\sum \frac{\ell_i}{k_i} + \frac{1}{h}} = 200$$

$$200 = \frac{1400 - 85}{\frac{1}{12}\left(\frac{9}{0.8} + \frac{4.5}{0.15} + \frac{9}{0.4}\right) + \frac{1}{h}}$$

$h = 0.792$ Btu/hr ft^2 °F

[2] fireclay:

$$\frac{q}{A} = 250 = \frac{\Delta t}{L/k} = \frac{2000-1800}{L/0.9}$$

L = 0.72 ft = 8.64 inches = 2 bricks

$$\frac{q}{A} = 250 = \frac{t_1 - t_2}{\frac{1}{12}\left(\frac{9}{0.9}\right)} = \frac{2000 - t_2}{0.833}$$

$t_2 = 1791.7$°F

insulating:

$$250 = \frac{1791.7 - 300}{L_2/K_2} = \frac{1491.7}{L_2/0.12}$$

$L_2 = 0.716$ ft = 8.59 inches = 3 bricks

$$250 = \frac{1791.7 - t_3}{\frac{1}{12}(9/.12)}; \quad t_3 = 229.2\text{°F}$$

building

$$250 = \frac{229.2 - 100}{L_3/K_3} = \frac{129.2}{L_3/0.4}$$

$L_3 = .207$ ft = 1 brick

[a]
total thickness = (9 + 9 + 4) = 22″
(2 fireclay + 3 insulating + 1 building)

[b]
$$\frac{q}{A} = \frac{2000 - 100}{\frac{1}{12}\left(\frac{9}{.9} + \frac{9}{.12} + \frac{4}{.4}\right)} = 240 \frac{\text{Btu}}{\text{hr ft}^2}$$

[3]

$$q = \frac{4\pi k \Delta T r_i r_o}{r_i - r_o}$$

$r_i = \frac{3}{12}$ ft ; $r_o = \frac{5}{12}$ ft

$$q = \frac{4\pi(8)(212-300)\left(\frac{3}{12}\right)\left(\frac{5}{12}\right)}{\left(\frac{3}{12} - \frac{5}{12}\right)}$$

$q = 5529$ Btu/hr

[4] ignoring edge and corner losses:
for thickness x:

$$q = \frac{\Delta t}{R}; \quad R = \frac{x}{KA}; \quad A = 2[(5)(4) + 5(3) + 4(3)]$$

$A = 94$ ft^3

but

$$q = \frac{1440000}{24} = 60000 \text{ Btu/hr}$$

$$R = \frac{500 - 100}{60000} = 0.665 \times 10^{-2} \frac{\text{°F hr}}{\text{Btu}}$$

$x = RKA = 0.665 \times 10^{-2} \times 0.8 \times 94$

x = 0.5 ft = 6 inches

[5]

$$q = \frac{T_1 - T_3}{\frac{1}{2\pi}\left[\frac{\ln r_2/r_1}{K_1} + \frac{\ln r_3/r_2}{K_2}\right]}$$

$K_{asbestos} = 0.12$ Btu/hr-ft-°F
$K_{cork} = 0.03$ Btu/hr-ft°F

$r_1 = 1$ ft ; $r_2 = \frac{14}{12}$ ft ; $r_3 = \frac{15.5}{12}$ ft

$$q = \frac{300 - 85}{\frac{1}{2\pi(1)}\left[\frac{\ln 14/12}{.12} + \frac{\ln 15.5/14}{.03}\right]}$$

$q = 288.8$ Btu/hr / ft of pipe

[6] Heat transfer by radiation & natural convection.

$$q_r = A\sigma(T_1^4 - T_2^4)\frac{1}{\frac{1}{\varepsilon_1}+\frac{1}{\varepsilon_2}-1}$$

$$q_r = 150(.1713\times10^{-8})(530^4-460^4)\frac{1}{\frac{1}{.9}+\frac{1}{.9}-1} = 7180$$

$$q_c = h_c(T_1 - T_2)$$

$$\frac{h_c X}{K_f} = \left[\frac{X_p f^2 g \beta_f \Delta t}{\mu_f^2}\ \frac{C_p\mu}{k}\right]^m \frac{C_*}{(L/X)^{1/9}}$$

(1) assume no Al, $\varepsilon = 0.9$ for wood

air properties at 35°F

$C_p = 0.240$ Btu/lb°F

$\mu = 0.042$ lb/ft-hr

$K = 0.0141$ Btu/hr-ft°F

$$\frac{C_p\mu}{K} = 0.72 = N_{Pr}$$

let $Y = \frac{\rho_f^2 \beta_f g C_p}{\mu_f k_f} = 2.16\times10^6/\text{ft-°F}$

$C_* = 0.071$ & $m = 1/3$ when $N_{Gr} = 2.1\times10^5$ to 1.1×10^7

$$\frac{h_c(1/3)}{0.0141} = \frac{0.071}{[10/(1/3)]^{1/9}}\left[Y\Delta t X^3\right]^{1/3}$$

$$= \frac{0.071}{(30)^{1/9}}\left[\frac{2.16\times10^6(70)}{27}\right]^{1/3}$$

$h_c = 0.3645$

$q_c = (0.3645)(150)(70) = 3828$ Btu/hr

$q = q_r + q_c = 7180 + 3828 = 11008$ Btu/hr

(2) with Al present
since properties of air not much different on either side of foil.

$\varepsilon_{Al} = 0.087$

$$q_r = (150)(.1713\times10^{-8})(530^4-495^4)\left[\frac{1}{\frac{1}{.9}+\frac{1}{.087}-1}\right] = 418\ \frac{\text{Btu}}{\text{hr}}$$

$$N_{Gr} = \frac{Yx^3\Delta t}{N_{Pr}} = \frac{1.92\times10^6(1/6)^3(35)}{0.72}$$

$= 4.32\times10^5 \Rightarrow C = 0.071;\ m = \frac{1}{3}$

$$\frac{h_c X}{K_f} = \frac{0.071}{(60)^{1/9}}\left[4.32\times10^5 \times .72\right]^{1/3}$$

$h_c = 0.265;\quad q_c = 150(.267)(35) = 1402$

$q = h_c + h_r = 1402 + 418 = 1820$ Btu/hr

$$\frac{q_{Al}}{q_{without}} = \frac{1820}{11008} = 0.165$$

reduction by 84% by using Al.

[7] $Y = \frac{t'-t}{t'-t_b} = \frac{60-100}{60-300} = 0.1666$

$\eta = \frac{r}{r_m} = \frac{0}{1/12} = 0$

$m = \frac{k}{r_m h} = 0$ since $h \gg \frac{k}{r_m}$

$X = 0.83 = \frac{\alpha\theta}{r_m^2}$; $\alpha = \frac{k}{\rho C_p} = \frac{.4}{155(.2)} = .0129$

$$\theta = \frac{(0.83)r_m^2}{\alpha} = \frac{(0.83)(\frac{1}{12})^2}{0.0129}$$

X determined from figure 6.7

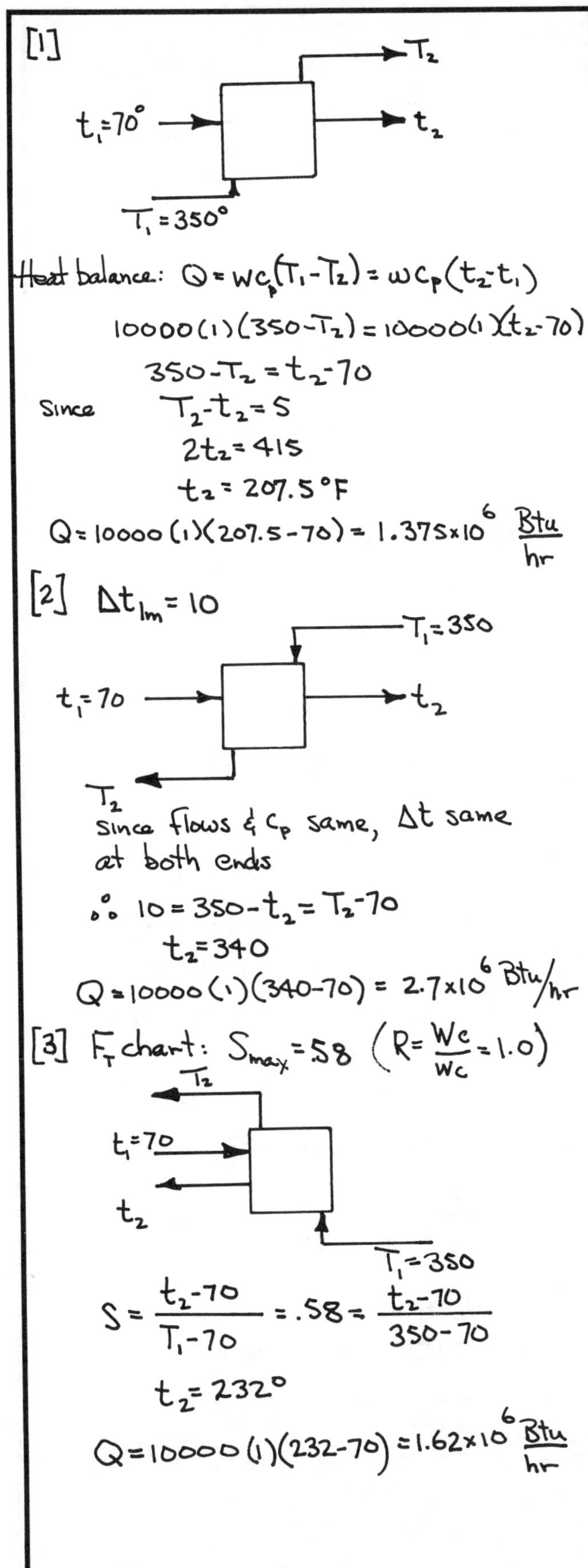

[1]

$t_1 = 70°$, $T_1 = 350°$, outputs T_2, t_2

Heat balance: $Q = wc_p(T_1 - T_2) = wc_p(t_2 - t_1)$

$10000(1)(350 - T_2) = 10000(1)(t_2 - 70)$

$350 - T_2 = t_2 - 70$

Since $T_2 - t_2 = 5$

$2t_2 = 415$

$t_2 = 207.5°F$

$Q = 10000(1)(207.5 - 70) = 1.375 \times 10^6 \ \frac{\text{Btu}}{\text{hr}}$

[2] $\Delta t_{lm} = 10$

$t_1 = 70$, $T_1 = 350$, outputs t_2, T_2

Since flows & c_p same, Δt same at both ends

$\therefore 10 = 350 - t_2 = T_2 - 70$

$t_2 = 340$

$Q = 10000(1)(340 - 70) = 2.7 \times 10^6 \ \text{Btu/hr}$

[3] F_T chart: $S_{max} = .58 \ \left(R = \frac{Wc}{wc} = 1.0\right)$

$t_1 = 70$, $T_1 = 350$, outputs T_2, t_2

$$S = \frac{t_2 - 70}{T_1 - 70} = .58 = \frac{t_2 - 70}{350 - 70}$$

$t_2 = 232°$

$Q = 10000(1)(232 - 70) = 1.62 \times 10^6 \ \frac{\text{Btu}}{\text{hr}}$

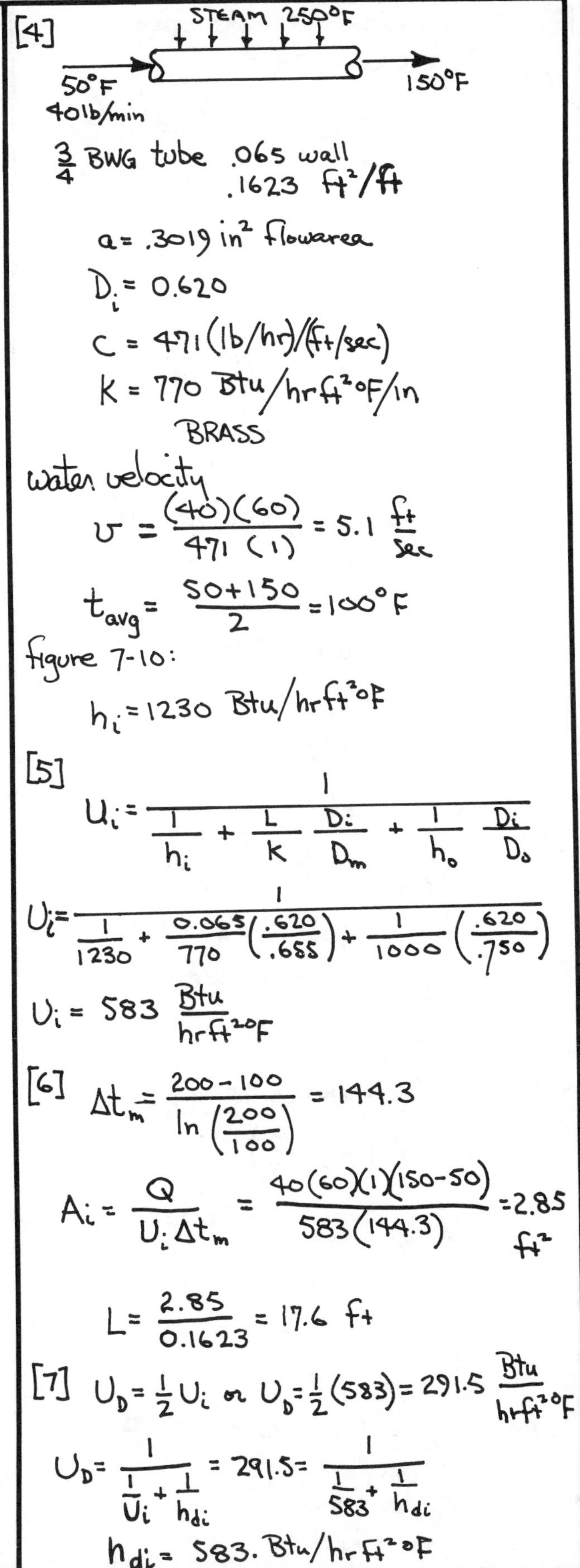

[4]

STEAM 250°F

50°F, 40 lb/min → 150°F

$\frac{3}{4}$ BWG tube .065 wall

.1623 ft^2/ft

$a = .3019 \ \text{in}^2$ flowarea

$D_i = 0.620$

$C = 471 \ (\text{lb/hr})/(\text{ft/sec})$

$K = 770 \ \text{Btu/hr ft}^2 \text{°F/in}$

BRASS

water velocity

$$v = \frac{(40)(60)}{471\ (1)} = 5.1 \ \frac{\text{ft}}{\text{sec}}$$

$$t_{avg} = \frac{50 + 150}{2} = 100°F$$

figure 7-10:

$h_i = 1230 \ \text{Btu/hr ft}^2 \text{°F}$

[5]

$$U_i = \frac{1}{\frac{1}{h_i} + \frac{L}{K}\frac{D_i}{D_m} + \frac{1}{h_o}\frac{D_i}{D_o}}$$

$$U_i = \frac{1}{\frac{1}{1230} + \frac{0.065}{770}\left(\frac{.620}{.685}\right) + \frac{1}{1000}\left(\frac{.620}{.750}\right)}$$

$U_i = 583 \ \frac{\text{Btu}}{\text{hr ft}^2 \text{°F}}$

[6]

$$\Delta t_m = \frac{200 - 100}{\ln\left(\frac{200}{100}\right)} = 144.3$$

$$A_i = \frac{Q}{U_i \Delta t_m} = \frac{40(60)(1)(150 - 50)}{583(144.3)} = 2.85 \ \text{ft}^2$$

$$L = \frac{2.85}{0.1623} = 17.6 \ \text{ft}$$

[7] $U_D = \frac{1}{2}U_i$ or $U_D = \frac{1}{2}(583) = 291.5 \ \frac{\text{Btu}}{\text{hr ft}^2 \text{°F}}$

$$U_D = \frac{1}{\frac{1}{U_i} + \frac{1}{h_{di}}} = 291.5 = \frac{1}{\frac{1}{583} + \frac{1}{h_{di}}}$$

$h_{di} = 583. \ \text{Btu/hr ft}^2 \text{°F}$

[8] This problem depends upon initial assumptions

$$q_{in} = (q_{out})_{radiation} + (q_{out})_{convection}$$

ASSUMPTION 1: RADIATION LOSS TO AIR AT 90°

$\varepsilon = 1$

$$350 = (.173)(1)\left[\left(\frac{t_s+460}{100}\right)^4 - \left(\frac{550}{100}\right)^4\right] + 2.8(t_s - 90)$$

$t_s = 172.3°F$ (using bisection method)

ASSUMPTION 2: RADIATION LOSS TO SPACE @ 0°R, $\varepsilon = 1$

$$350 = .173(1)\left[\left(\frac{t_s+460}{100}\right)^4 - 0\right] + 2.8(t_s - 90)$$

$t_s = 136°F$

ASSUMPTION 3: No radiation loss

$$350 = 2.8(t_s - 90)$$

$t_s = 215°F$

[9] Enthalpy balance

(a)

$$W_{H_2O} C_p (t_o - t_i) = \Delta h_{vap} W_{CH_3OH}$$

$$W_{H_2O}(1)(115-70) = (600)(10000)$$

$$W_{H_2O} = 133\,333 \text{ lb/hr}$$

(b)

$$q = UA\Delta t_{lm}$$

$$\Delta t_{lm} = \frac{(160-70)-(160-115)}{\ln\left(\frac{160-70}{160-115}\right)} = 65°$$

$$q = \Delta H_{vap} W_{CH_3OH} = 600(10000)$$

$$q = 6\times10^6 \ \frac{\text{Btu}}{\text{hr}}$$

$$6\times10^6 = (400)A(65)$$

$$A = 231 \text{ ft}^2$$

(c) no heat loss to surroundings
U constant
C_p constant
flow steady

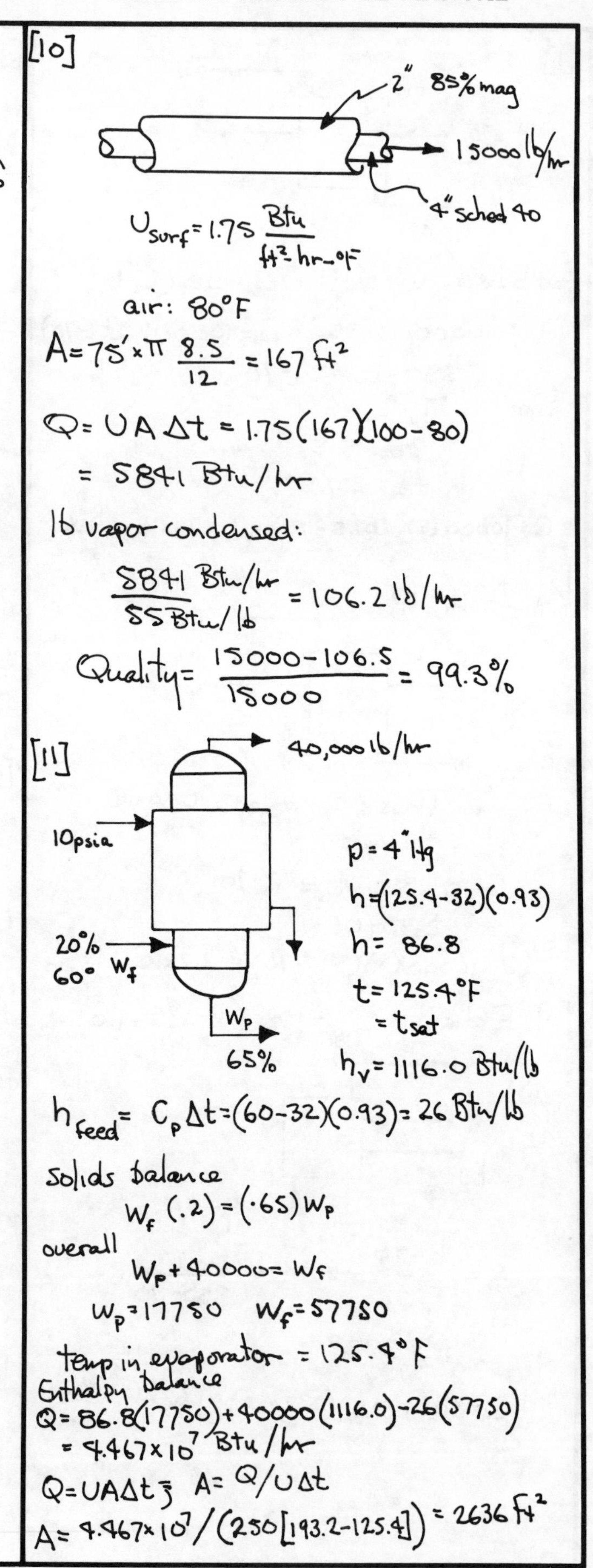

[10]

$U_{surf} = 1.75 \ \frac{\text{Btu}}{\text{ft}^2\text{-hr-°F}}$

air: 80°F

$$A = 75' \times \pi \ \frac{8.5}{12} = 167 \text{ ft}^2$$

$$Q = UA\Delta t = 1.75(167)(100-80) = 5841 \text{ Btu/hr}$$

lb vapor condensed:

$$\frac{5841 \text{ Btu/hr}}{55 \text{ Btu/lb}} = 106.2 \text{ lb/hr}$$

$$\text{Quality} = \frac{15000 - 106.5}{15000} = 99.3\%$$

[11]

$p = 4"$ Hg

$h = (125.4-32)(0.93)$

$h = 86.8$

$t = 125.4°F = t_{sat}$

$h_v = 1116.0$ Btu/lb

$$h_{feed} = C_p\Delta t = (60-32)(0.93) = 26 \text{ Btu/lb}$$

solids balance

$$W_f(.2) = (.65)W_p$$

overall

$$W_p + 40000 = W_f$$

$$W_p = 17750 \quad W_f = 57750$$

temp in evaporator = 125.4°F

Enthalpy balance

$$Q = 86.8(17750) + 40000(1116.0) - 26(57750) = 4.467\times10^7 \text{ Btu/hr}$$

$$Q = UA\Delta t; \quad A = Q/U\Delta t$$

$$A = 4.467\times10^7/(250[193.2-125.4]) = 2636 \text{ ft}^2$$

[11] con't

steam req'd: $\frac{Q}{\Delta h_v} = S$

$S = 4.4679 \times 10^7 / 982.1$

$S = 45500$ lb/hr steam

[12]
(a) an approximate calculation can show the effect of oil inside vs. outside

oil outside:

$$\frac{1}{U_o} \approx \frac{1}{h_o} + \frac{1}{h_i} = \frac{1}{41} + \frac{1}{2000}$$
$$= .0249$$

oil inside

$$\frac{1}{U_o} \approx \frac{1}{h_o} + \frac{1}{h_i} = \frac{1}{90} + \frac{1}{840}$$
$$= .0123$$

U_o is about twice as large for oil on inside. Since oil is controlling fluid, it should be on outside where U is highest

(b) 1" sched 40 $D_i = 1.049''$ $D_o = 1.315''$

$\pi D_o = 0.344$ ft wall $= X_w = .133''$

$K_m = 26$ Btu-ft/hr°F ft²

$$\frac{1}{U_o} = \frac{1}{h_o} + \frac{X_w D_o}{K D_i} + \frac{D_o}{h_i D_i} = \frac{1}{840} + \frac{.133(1.315)}{12(1.18)} \frac{1}{26} + \frac{1.315}{90(1.049)}$$

$U_o = 64$ Btu/ft²°F-hr

$q = UA\Delta t = wC_p(t_i - t_o) = 800(.4)(150-40)(1.8)$

$q = 633600$ Btu/hr

$$\Delta t = \frac{(150-40)-(40-20)}{\ln \frac{110}{20}} = 52.8°C$$

$U_o A \Delta t = U \pi D_o L \Delta t = q$

$$L = \frac{q}{U \pi D_o \Delta t} = \frac{63360}{64(.344)(52.8)(1.8)}$$

$L = 303$ ft

[13] Assume air film controlling $h = U$, and 1 means first case. From Sieder Tate equation it can be shown if other fluid properties constant:

$U_1 = K V_1^{0.8}$

$U_2 = K V_2^{0.8}$

$q_1 = w_1 C_p(t_2 - t_1) = U_1 A \Delta t_1$

$q_2 = w_2 C_p(t_2 - t_1) = U_2 A \Delta t_2$

$$\Delta t_1 = \frac{(220-80)-(220-180)}{\ln 140/40} = 79.8°$$

$$\Delta t_2 = \frac{(250-80)-(250-170)}{\ln 170/70} = 113°$$

W is proportional to velocity

$$\frac{W_2}{W_1} = \frac{W_2^{0.8} \Delta t_2}{W_1^{0.8} \Delta t_1} \text{ or } \frac{W_2^{0.2}}{W_1^{0.2}} = \frac{\Delta t_2}{\Delta t_1} = 1.416$$

$W_2 = W_1 (1.416)^5 = W_1 (5.7)$

Approximately 5.7 times as much air can be heated with higher temp. steam.

[14]

From tubing characteristics

$D_o = 1.5''$ $D_i = 1.37''$ $X_w = 0.065$

$A_i = 0.0102\ ft^2$ $h_o = 1300\ Btu/ft^2hr°F$

$h_i = 600\ Btu/ft^2hr°F$

$k_m = 105\ Btu\text{-}in/ft^2hr°F$

Assume design for 6 ft/sec

$V = 6\ ft/sec;\ \rho = (1.3)62.4 = 81.2\ lb/ft^3$

$W = 650\,000\ lb/hr$

m = number of passes

η = number of tubes

$\eta A_i \bar{v}\rho = W$

$$\eta = \frac{650\,000}{(.0102)(6)(3600)(81.12)} = 36.4$$

$\eta = 36$

$$\frac{1}{U_o} = \frac{1}{h_o} + \frac{D_o X_w}{\bar{D}\ k_m} + \frac{D_o}{h_i D_i}$$

$$\bar{D} = \frac{1.50 - 1.37}{\ln \frac{1.5}{1.37}} = 1.43$$

$$\frac{1}{U_o} = \frac{1}{1300} + \frac{(1.5)(.065)}{(1.43)(105)} + \frac{1.5}{1.37(600)}$$

$U_o = 308\ Btu/ft^2\text{-}hr°F$

$A_o = n\pi D_o L m = (169.7)m$

$q = c_p \Delta t W = U_o A_o \Delta t_{lm}$

$$\Delta t_{lm} = \frac{(t_s - 148) - (t_s - 198)}{\ln \frac{t_s - 148}{t_s - 198}}$$

$t_s = 258.8;\ \Delta t_{lm} = 83.5°F$

$q = (0.78)(198 - 148)(650000) = 308(169.7m)83.5$

$m = 5.82$

use 6 passes of 36 tubes/pass

[15]

cocurrent:

$Q_{oil} = m_{oil} C_p (t_o - t_i) = 2000(.56)(200 - 60)$

$= 123200\ Btu/hr$

$Q_{kero} = 123200\ Btu/hr$

$$m_{kero} = \frac{123200}{(0.6)(450 - 220)} = 893\ lb/hr$$

countercurrent:

$123200 = m(0.6)(450 - 110)$

$m = 604\ lb/hr$

∴ flow rate greater for cocurrent

Compare at same kero flow, 893 lb/hr

$$(\Delta t_{lm})_{cocurrent} = \frac{360 - 20}{\ln \frac{360}{20}} = 118°F$$

$$(\Delta t_{lm})_{counter} = \frac{250 - 130}{\ln \frac{250}{130}} = 183°F$$

where Q & U are fixed counter-current requires less area at same flowrate.

at min flow rates:

$$(\Delta t_{lm})_{counter} = \frac{250 - 20}{\ln \frac{250}{20}} = 91.2°F$$

at min flowrates countercurrent needs more area.

[16] $\frac{1}{U_1} = \frac{1}{U_c} + \frac{t}{k} + \frac{1}{U_{\ell_1}} = \frac{1}{250} = \frac{1}{1200} + \frac{.12}{460} + \frac{1}{U_{\ell_1}}$

$U_{\ell_1} = 344$

Area $1\frac{1}{2}$ tube 0.12 wall = .00865 ft^2
0.065 wall = .0102 ft^2

$V_2 = V_1 \frac{.00865}{.0102} = .845\, V_1$

$U_{\ell_2} = U_{\ell_1}\sqrt{\frac{.845 V_1}{V_1}} = 316$

$\frac{1}{U_2} = \frac{1}{1200} + \frac{0.065}{105} + \frac{1}{316}; \ 217 \frac{Btu}{hr\,ft^2\,{}^\circ F} = U_2$

A PCU is the heat req'd to raise 1 lb of H_2O 1°C

1 PCU = 1.8 Btu

[17] $q = UA\Delta t_{lm}$

$\Delta t_{lm} = \frac{(80.1-35)-(80.1-65)}{\ln 45.1/15.1} = 27.5°C = 49.5°F$

$A_o = 36(18)(.2618) = 169.5\ ft^2$

$q = 5000(170) = 8.5\times10^5$ Btu/hr

$U_o = \frac{q}{A_o \Delta t_{lm}} = \frac{8.5\times10^5}{(169.5)49.5} = 101 \frac{Btu}{hr\,ft^2\,{}^\circ F}$

$\frac{1}{U_o} = \frac{D_o}{D_i h_i} + \frac{x_w D_o}{k_m D_L} + \frac{1}{h_o}; \ \frac{1}{101} = \frac{1.0}{.87 h_i} + \frac{.065(1.0)}{12(9.4)(.94)} + \frac{1}{300}$

$h_i = 193 \frac{Btu}{hr\,ft^2\,{}^\circ F}$

For heat transfer in tubes

$h_i = .023 \frac{K}{D_i}(N_{Re})^{.8}(N_{Pr})^{1/3}$

$\left(\frac{C_p \mu}{K}\right)^{1/3} = \left[\frac{(1.0)(0.56)(2.42)}{0.372}\right]^{1/3} = 1.54$

H_2O rate $= \frac{q}{C_p \Delta t} = \frac{8.5\times10^5}{(1.0)30(1.8)} = 1.57\times10^4 \frac{lb}{hr}$

Cross section: $36(.00413) = 0.1485\ ft^2$

$G = 1.57\times10^4/.1485 = 1.06\times10^5\ lb/hr\text{-}ft^2$

$N_{Re} = \frac{DG}{\mu} = \frac{(0.87/12)(1.06\times10^5)}{(.56)(2.42)} = 5650$

$N_{Re}^{.8} = 1000$

$h_i = \frac{(0.023)(.372)(1000)(1.54)}{.87/12}$

$h_i = 182\ Btu/hr\,ft^2\,{}^\circ F$

Since h_i was actually 193 heat exchanger is performing properly with no evidence of fouling.

[18]

Enthalpy balance:

	Aniline	Toluene
t_{avg}	125°	
C_p	0.52	0.46

Aniline:

$$Q = wC_p\Delta t = 7000(.52)(150-100)$$
$$= 182000 \text{ Btu/hr}$$

Toluene:

$$\Delta t = \frac{Q}{wC_p} = \frac{182000}{(10000).46} = 39.5°F$$

$$T_{out} = 185 - 39.5 = 145.5°F$$

$$\Delta t_{lm} = \frac{45.5 - 35}{\ln \frac{45.5}{35}} = 40°F$$

Using following equations

D_e annulus $= D_2 - D_1$

$G = W/A$

$N_{Re} = DG/\mu$

$N_{Pr} = c_p\mu/k$

$h = 023 \frac{K}{D} N_{Re}^{.8} N_{Pr}^{1/3}$

hot: annulus, toluene	cold: inner aniline
$D_e = 0.625'$	$D_e = .0874'$
$G = 719000 \frac{\text{lb}}{\text{ft}^2\text{-hr}}$	$G = 1162000$
$\mu = 0.848$ lb/ft-hr	$\mu = 4.72$
$N_{Re} = 5.3 \times 10^4$	$N_{Re} = 2.51 \times 10^4$
$N_{Pr} = 4.63$	$N_{Pr} = 25$
$h_o = 340$	$h_o = 176$

$$\frac{1}{U_o} = \frac{D_o}{D_i h_i} + \frac{X_W D_o}{K_m D_L} + \frac{1}{h_o} + \frac{1}{h_{DIRTY}}$$

$$\frac{X_W D_o}{K_m D_L} \approx \frac{1}{26} \frac{.133}{12} \frac{1.315}{1.182} = 0.47 \times 10^{-3}$$

$$\frac{1}{U_o} = \frac{1}{176} + 0.47 \times 10^{-3} + \frac{1}{340} + 5 \times 10^{-3}$$

$$\frac{1}{U_o} = .014092$$

$U_o = 71$ $\quad Q = U_o A_o \Delta t$

$$A_o = \frac{182000}{(71)(40)} = 64.1 \text{ ft}^2$$

$$L = \frac{64.1 \text{ ft}^2}{.344 \text{ ft}^2/\text{ft}} = 186 \text{ ft}$$

EACH HAIRPIN = 30 ft

$\therefore \frac{186}{30} = 6$ hairpins

[1] $\sum p_i = \sum p_i^o x_i = 3000 = 14000 x_E + 2700 x_P + 500 x_B$

since $x_P = x_B$; ∴ $x_E + x_P + x_B = 1$

$x_P = x_B = \frac{1-x_E}{2}$

$3000 = 14000 x_E + \frac{1-x_E}{2}(2700+500)$

$x_E = 0.1129$; $x_P = .4435$; $x_B = .4435$

$y_i = \frac{p_i}{p} = \frac{p_i^o x_i}{p}$

$y_E = 14000(.1129)/3000 = .5269$

$y_P = 2700(.4435)/3000 = .3992$

$y_B = 500(.4435)/3000 = .0739$

[2] min H_2O rate occurs when bottoms is in equilibrium with feed

$MW_{air} = 29$; $MW_{acetone} = 58$

$MW_{mixture} = .02(58) + .98(29) = 29.58$

acetone in feed $= .02\left(\frac{1000}{29.58}\right) = .676$ lbmole/hr

acetone in bottoms $= .95(.676) = .642$ lbmole/hr

from equilibrium: $M_A = 2.53 x_a = 2.0$ mole %

$x_a = 2/2.53 = 0.791$ mole %

$x_{H_2O} = 100 - .791 = 99.209$ mole %

H_2O in bottoms $= 99.209(.642)/.791$

$= 80.52$ lbmole/hr

$80.52(18) = 1449$ lb H_2O/hr

[3] ASSUME EQUIMOLAR IS IN GAS PHASE

A & B are organic acids

$x_A p_A^o + x_B p_B^o = 200 - p_{H_2O}$

$y_A = y_B$ & $x_A + x_B = 1$

∴ $x_A p_A^o = x_B p_B^o = (1 - x_A) p_B^o$

$x_A = p_B^o/(p_A^o + p_B^o) = 32/(14+32)$

$x_A = .696$; $x_B = .304$

∴ $.696(14) + .304(32) = 200 - p_{H_2O}$

$p_{H_2O} = 180.52$; $y_{H_2O} = \frac{180.52}{200} = 0.9026$

$y_A = y_B = .696(14)/200 = .0487$

$\frac{\text{moles } H_2O}{\text{mole acid}} = \frac{0.9026}{2(.0487)}$

$\frac{\text{lb } H_2O}{\text{mole acid}} = \frac{0.9026}{2(.0487)}(18) = 167 \frac{\text{lb } H_2O}{\text{mole acid}}$

[4] molal diffusivity $= \mathcal{D}_m = \mathcal{D}\rho = 0.577 \times 10^{-3}$ lbmole/ft-hr

since ideal gas: $\rho = \frac{P}{RT}$; ∴ $\mathcal{D} = \mathcal{D}_m RT/P$

$N_A = \frac{P}{RT}\mathcal{D}\frac{\Delta p_A}{\Delta z (p_B)_{lm}} = \mathcal{D}_m \frac{\Delta p_A}{\Delta z (p_B)_{lm}}$

$\Delta p_A = \frac{20}{760} - 0 = .026315$ atm

$p_{B_1} = 1 - .026315 = .97368$ atm

$p_{B_2} = 1$ atm

$(p_B)_{lm} = \frac{1 - .97368}{\ln \frac{1}{.97368}} = 0.9866$ atm

$\Delta z = 5$ ft

$N_A = \frac{0.577 \times 10^{-3}(.026315)}{5(.9866)} = 3.078 \times 10^{-6} \frac{\text{lbmole}}{\text{ft}^2\text{-hr}}$

$N_A A = \frac{\pi}{4}\left(\frac{2}{12}\right)^2 3.078 \times 10^{-6} = 6.715 \times 10^{-8} \frac{\text{lbmole}}{\text{hr}}$

1% volume ≡ .01 mole fraction

$\ln\left(\frac{1-0.01}{1-0.026315}\right) = \frac{h}{\Delta z}\left\{\ln\left(\frac{1-0}{1-0.026315}\right)\right\}$

$h = 1.275$

$5 - h = 3.725$ ft from end of pipe

[5] $x_i = \frac{z_i}{K_i V + L}$; basis: 1 mole feed

i	z_i	K_i	x_i	y_i
CH_4	0.10	17.0	0.0111	0.1887
C_2H_6	0.20	3.1	0.0976	0.3026
C_3H_8	0.30	1.0	0.3000	0.3000
iC_4H_{10}	0.15	0.44	0.2083	0.0917
mC_4H_{10}	0.20	0.32	0.3030	0.0970
mC_5H_{12}	0.05	0.096	0.0912	0.0088
			1.0112	0.9888

Equil constants not consistent

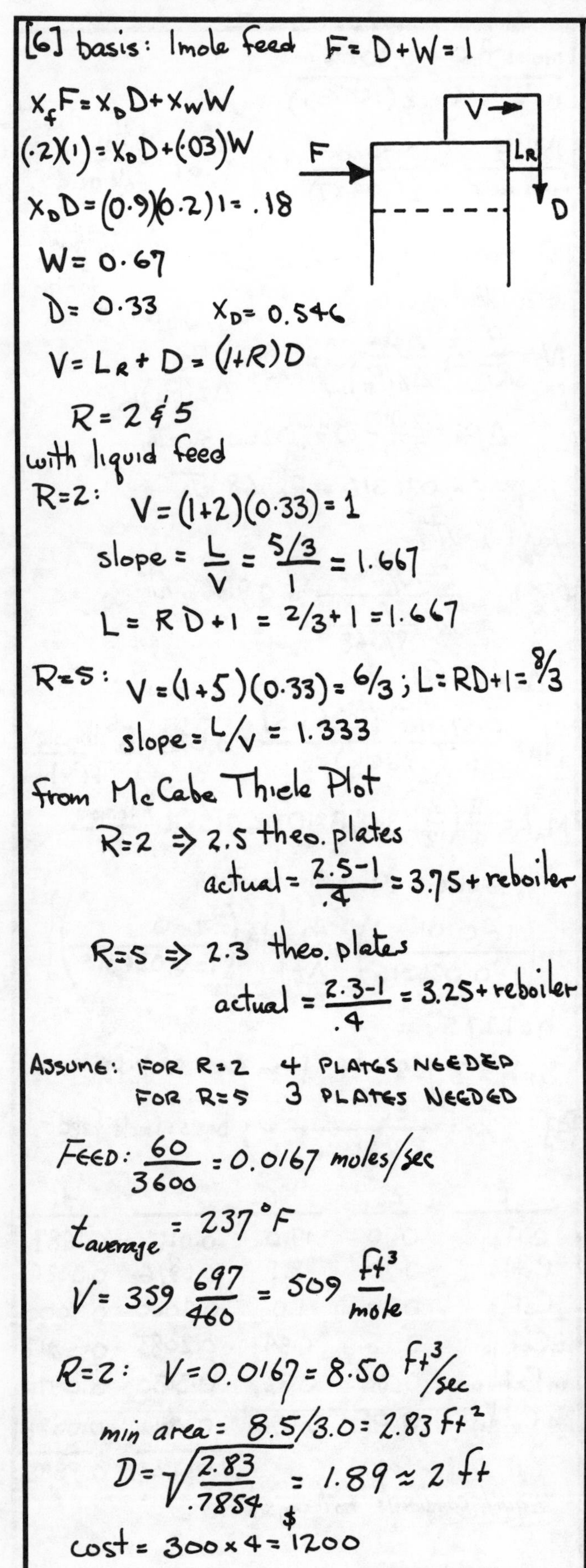

[6] basis: 1mole feed $F = D + W = 1$

$x_f F = x_D D + x_W W$

$(.2)(1) = x_D D + (.03)W$

$x_D D = (0.9)(0.2)1 = .18$

$W = 0.67$

$D = 0.33 \qquad x_D = 0.546$

$V = L_R + D = (1+R)D$

$R = 2 \,\&\, 5$

with liquid feed

R=2: $V = (1+2)(0.33) = 1$

$\text{slope} = \frac{L}{V} = \frac{5/3}{1} = 1.667$

$L = RD + 1 = 2/3 + 1 = 1.667$

R=5: $V = (1+5)(0.33) = 6/3 \,;\, L = RD + 1 = 8/3$

$\text{slope} = L/V = 1.333$

from McCabe Thiele Plot

R=2 ⇒ 2.5 theo. plates

$\text{actual} = \frac{2.5-1}{.4} = 3.75 + \text{reboiler}$

R=5 ⇒ 2.3 theo. plates

$\text{actual} = \frac{2.3-1}{.4} = 3.25 + \text{reboiler}$

ASSUME: FOR R=2 4 PLATES NEEDED
FOR R=5 3 PLATES NEEDED

FEED: $\frac{60}{3600} = 0.0167$ moles/sec

$t_{average} = 237°F$

$V = 359 \frac{697}{460} = 509 \frac{ft^3}{mole}$

R=2: $V = 0.0167 = 8.50\ ft^3/sec$

min area $= 8.5/3.0 = 2.83\ ft^2$

$D = \sqrt{\frac{2.83}{.7854}} = 1.89 \approx 2\ ft$

cost $= 300 \times 4 = \$1200$

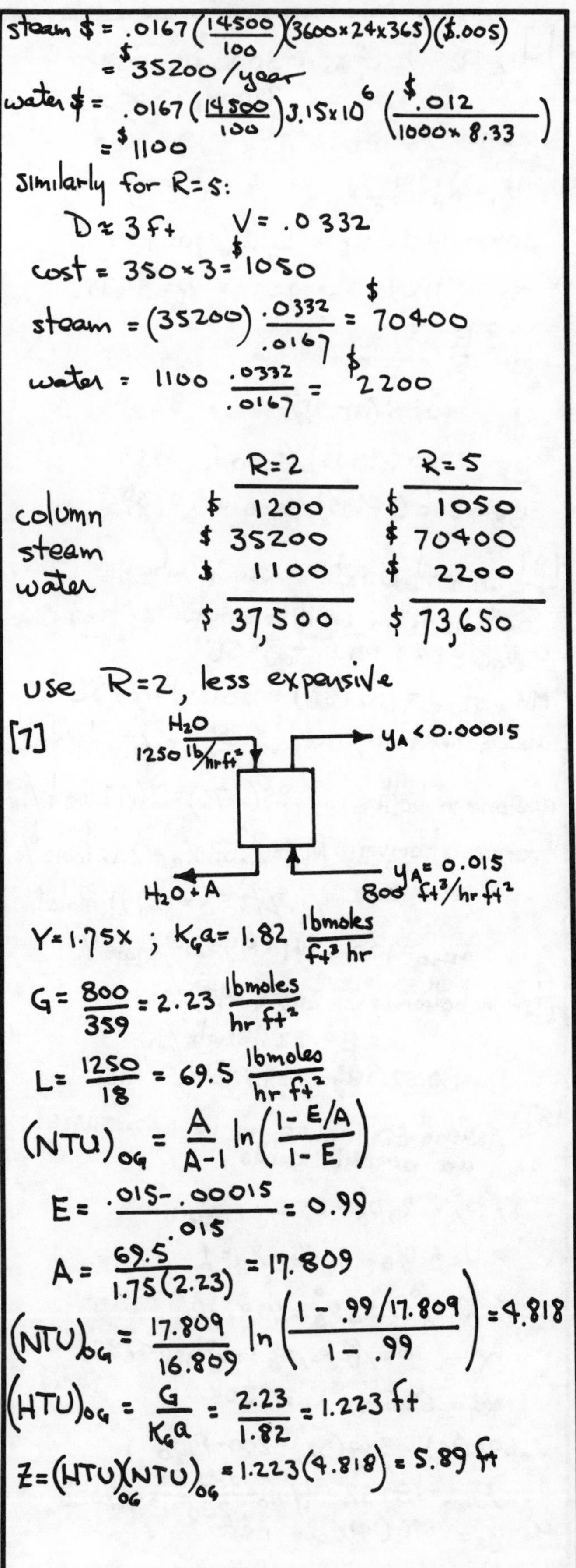

steam \$ $= .0167\left(\frac{14500}{100}\right)(3600 \times 24 \times 365)(\$.005)$

$= \$35200/\text{year}$

water \$ $= .0167\left(\frac{14500}{100}\right)3.15 \times 10^6 \left(\frac{\$.012}{1000 \times 8.33}\right)$

$= \$1100$

similarly for R=5:

$D \approx 3\ ft \qquad V = .0332$

cost $= 350 \times 3 = \$1050$

steam $= (35200)\frac{.0332}{.0167} = \$70400$

water $= 1100 \frac{.0332}{.0167} = \$2200$

	R=2	R=5
column	\$ 1200	\$ 1050
steam	\$ 35200	\$ 70400
water	\$ 1100	\$ 2200
	\$ 37,500	\$ 73,650

use R=2, less expensive

[7]

$Y = 1.75X \,;\, K_G a = 1.82 \frac{\text{lbmoles}}{ft^3\ hr}$

$G = \frac{800}{359} = 2.23 \frac{\text{lbmoles}}{hr\ ft^2}$

$L = \frac{1250}{18} = 69.5 \frac{\text{lbmoles}}{hr\ ft^2}$

$(NTU)_{OG} = \frac{A}{A-1} \ln\left(\frac{1 - E/A}{1 - E}\right)$

$E = \frac{.015 - .00015}{.015} = 0.99$

$A = \frac{69.5}{1.75(2.23)} = 17.809$

$(NTU)_{OG} = \frac{17.809}{16.809} \ln\left(\frac{1 - .99/17.809}{1 - .99}\right) = 4.818$

$(HTU)_{OG} = \frac{G}{K_G a} = \frac{2.23}{1.82} = 1.223\ ft$

$Z = (HTU)_{OG}(NTU)_{OG} = 1.223(4.818) = 5.89\ ft$

[8] Assume H_2O condenses first

$\bar{p}^{\circ}_{H_2O} = 0.3$ atm $= 4.41$ psia

(steam tables @ 156.96°F (= 616.6°R)

note "6" = hexane "7" = heptane

$y_6 = P_6^{\circ} x_6 \qquad y_6 = .6 \quad y_7 = .1 \quad y_s = .3$

$$\ln P_6^{\circ} = 17.719 - \frac{6816.4}{616.6} = 6.6566$$

$P_6^{\circ} = 778$ mm Hg

$$\ln P_7^{\circ} = 17.9184 - \frac{7547.4}{616.6} = 5.6786$$

$P_7^{\circ} = 292.2$ mm Hg

$$X_6 + X_7 = \frac{(.6)(760)}{778} + \frac{0.1\,(760)}{292.6} = 0.845$$

$\Sigma X_i < 1$ ∴ no condensation of hydrocarbons occurs at 156.96°F pure water condenses.

at $t = 150°F$ (= 609.7°R)

$P_s^{\circ} = 0.2530$ atm $= 192.28$ mm Hg

$$y_6 + y_7 = \frac{760 - 192.28}{760} = 0.747$$

since $y_6/y_7 = 6$

$y_6 = .6402 \qquad y_7 = .1067$

$\ln P_6^{\circ} = 6.5304$; $P_6^{\circ} = 685.7$ mm Hg

$\ln P_7^{\circ} = 5.5389$; $P_7^{\circ} = 254.4$ mm Hg

$$X_6 + X_7 = \frac{.6402\,(760)}{685.7} + \frac{.1067\,(760)}{254.4} = 1.028$$

by interpolation, $t = 151.1°F$

$X_6 = .7096/1.028 = .690$

$X_7 = .3188/1.028 = .3108$

since $X_6 + X_7 = .7096 + .3188 = 1.028$

[1]
WHILE THIS PROBLEM IS WORDED THE SAME AS PROB #3 CHAPTER 8, WE WILL ASSUME EQUIMOLAR ACIDS IN LIQUID PHASE FOR THIS PROBLEM.

$P_A = X_A P_A^\circ$; $X_c = .05$ $X_A = X_B = 0.475$

$\pi = \Sigma P_i = .475(32) + .475(14) + P_{H_2O} = 200$

$P_{H_2O} = 178.15$ mm Hg

$y_{H_2O} = 178.15/200 = .891$

$y_A + y_B = 1 - 0.891 = 0.109$

$$\frac{\text{lb } H_2O}{\text{mole acid}} = \frac{0.891(18)}{.109} = 146.76 \frac{\text{lb } H_2O}{\text{mole acid}}$$

[2] $537\text{ Kg} = 1183.9\text{ lb}$ MW $C_7H_{16} = 100$

MW $C_8H_{18} = 114$

"7" refers to C_7H_{16}

"8" refers to C_8H_{18}

$l_i = 1183.9(.5)(1/100 + 1/114) = 11.111$ moles

$$X_7 = \frac{1183.9(.5)(.01)}{1183.9(.5)(.01 + 1/114)} = 0.5327$$

$$-\ln \frac{l_f}{l_i} = \frac{1}{\alpha - 1}\left(\ln \frac{X_i}{X_f} - \alpha \ln \frac{1 - X_i}{1 - X_f}\right)$$

$$\ln \frac{11.111}{4.74} = 1\left(\ln \frac{.5327}{X_f} - 2 \ln \frac{.4673}{1 - X_f}\right)$$

reduces to:

$$.961 = \frac{(1 - X_f)^2}{X_f}$$

$X_f^2 - 2.961 X_f + 1 = 0$

$X_f = 0.389 = X_7$

$X_8 = 1 - .389 = 0.611$

$$y_7 = \frac{\alpha x}{1 + (\alpha - 1)x} = \frac{2(0.5327)}{1 + (1)(0.5327)} = .6951$$

$y_8 = 1 - .6951 = 0.3049$

When $\alpha = 1$ no enrichment takes place

$X_7 = .5327$ $X_8 = .4673$

[3] Fenske equation:

$$N_m = \frac{\ln\left[\left(\frac{.9}{.1}\right)\left(\frac{.9}{.1}\right)\right]}{\ln 2} - 1 = 5.34 \text{ or } 6 \text{ plates}$$

slope of op. line at min reflux calculated from line through (.9, .9) and (.5, y^*)

$$y^* = \frac{\alpha X_f}{1 + (\alpha - 1) X_f} = .667$$

$$\text{slope} = S = \frac{.9 - .667}{.9 - .5} = .58325$$

$$S = \frac{R_m}{1 + R_m} \text{ or } R_m = \frac{S}{1 - S} = \frac{.58325}{1 - .58325}$$

$R_m = 1.399$

$R = 1.2(1.399) = 1.679$

$$y = \frac{1.679}{1 + 1.679} X + \frac{1}{1 + 1.679} X_D = .627X + .336$$

Underwood

$.627 K^2 + (.627 + .336 - 2)K + .336 = 0$

$.627 K^2 - 1.037 K + .336 = 0$

$K^2 - 1.654 K + .536 = 0$

$K_1 = .442$ $K_2 = 1.212$

Since the feed line is vertical, it has equation $X = X_f$

the intersection of the rectifying op. line and feed line is (X_f, y_i)

$y_i = .627(.5) + .336 = 0.650$

stripping line slope

$$\left(\frac{L}{V}\right)_s = \frac{.65 - .1}{.5 - .1} = 1.375$$

strip line op.

$y = 1.375 X - .0375$

Underwood (strip)

$1.375 K^2 + (1.375 - .0375 - 2)K - .0375 = 0$

$1.375 K^2 - K - 0.0375 = 0$

[3] con't

$K^2 - .727 - .027 = 0$

$K_1 = -0.035 \; ; K_2 = .762$

rectifying:

$$n_R \ln\left(\frac{2/.627}{[1+(1).442]^2}\right) = \ln \frac{[(.9)-.442][1.212-.5]}{[.5-.442][1.212-.9]}$$

$n_R \; 0.428 = 2.892$

$n_R = 6.756$

stripping:

$$n_s \ln\left(\frac{2/1.375}{[1+(1)(0.035)]^2}\right) = \ln \frac{(.5+.035)(.762-.1)}{(.1+.035)(.762-.5)}$$

$n_R \; 0.446 = 2.304$

$n_R = 5.166$

13 stages: 7 rectifying; 6 stripping

[4] see graphical solution 9.4 on page 42

[5] the sample is a point on the op. line.

The op line passes through (.52, .61) and (.95, .95)

$$\frac{L}{V} = \frac{R}{R+1} = \frac{.95-.61}{.95-.52} = .791$$

$R = 3.785$

$$D = F\,\frac{x_F - x_B}{x_D - x_B} = 1000\,\frac{0.3-0.06}{0.95-0.06}$$

$D = 269.7$ lbmoles/hr

$V = D(R+1) = 269.7(4.785) = 1290.5 \; \frac{\text{lbmoles}}{\text{hr}}$

$\Delta H_{v\,benzene} = 94.14 \frac{cal}{g} \times 78.1\, g/gmole = 7354$

$\Delta H_{v\,toluene} = 86.8 \times 92.2 = 7999$

$\Delta H_{v\,mix} = 7354(.95) + 7999(.05) = 7386$ cal/gmole

$\Delta H_{v\,mix} = 7386(1.8) = 13313$ Btu/lbmole

$Q = 1290.5\,(13313) = 1.718 \times 10^7$ Btu/hr

Assume cooling water

In: 80°F Out 120°F

Assume b.p = 180°F

$\Delta t_1 = 180 - 80 = 100$

$\Delta t_2 = 180 - 120 = 60$

$\Delta t_{avg} = 80$

Assume $U = 300$

$Q = UA\Delta t$

$1.718 \times 10^7 = 300(A)\,80$

$A = 716 \text{ ft}^2$

A preheater can be used depending upon whether the reflux ratio, total vapor, reboiler rates are held constant. Preheating the feed would increase the vaporate in column, improving production rate while putting a higher load on the reboiler and condenser.

[6]

Normal hexane and n-heptane form an almost ideal solution and fit the criteria needed to apply the McCabe-Thiele method of equilibrium stage calculations. Thus, an x-y diagram is sufficient to evaluate the column operation.

a. Reducing the reflux ratio from infinity (total) reflux to 10 will cause the distillate and bottoms composition to move slightly towards the feed composition. This happens because there is less room on the x-y diagram to fit the fixed number of equilibrium stages in the column.

b. Reducing the reflux ratio to a value of 1 will give a slope of the upper operating line of 0.5. This will intersect the equilibrium line if it starts from the total reflux composition of $x_D = 0.98$. The column will operate at this reflux ratio, but the distillate composition will be greatly reduced to a lower value, perhaps $x_D = 0.7$, and the bottoms will contain more n-hexane.

c. Increasing the steam pressure will increase the boilup rate and tend to increase the reboiler reflux ratio. Consequently, the overhead composition, x_D, will increase, and the bottoms, x_B, will decrease (an increase in separation of the higher temperature boiling component). The increased vapor rate may cause flooding

[6] con't

or partial flooding of the trays and cause a lower tray efficiency, decreasing the overhead composition.

d. Increasing the liquid and vapor rates can be accomplished by increasing the feed rate, reboiler heat input, and condenser cooling rate. The latter two will increase overhead and decrease bottoms composition. Increasing the feed rate will increase vapor and liquid rates and cause flooding. The result will be a lowering of plate efficiency, negating the improvements gained by changing reboiler heat input and condenser cooling water rate.

[7]

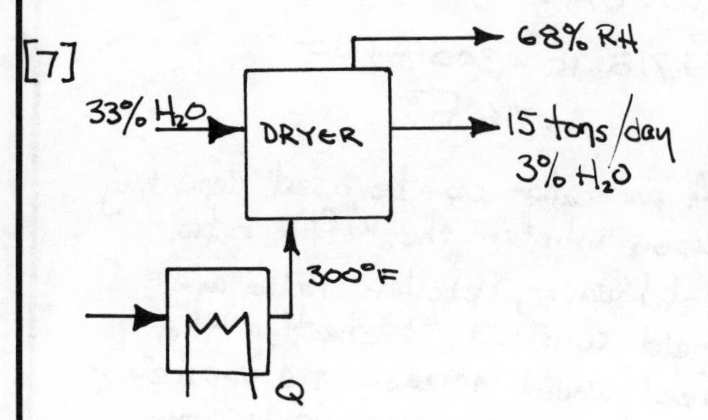

lb/day dry material = 0.97(30000)
= 29100 lb/day

lb/day feed = 29100/.67
= 43433 lb/day

68%
drying
45%
heating
.007
68°
300°

water removed = 43433 − 30000 = 13433 lb/day

from chart:

inlet humidity = .007 lb H_2O/lb dry air

heated air wet bulb = 105°F = outlet wet bulb

outlet temp = 116°F

outlet humidity = 0.048 lb H_2O/lb air

$$\text{lb dry air/hr} = \frac{13433/24}{(.048 - .007)} = 13659 \text{ lb/hr}$$

specific volume of air @ 45% RH, 68°F

13.4 ft³/lb dry air

$$\text{air input} = 13.4 \frac{\text{ft}^3}{\text{lb air}} \left(13659 \frac{\text{lb air}}{\text{hr}}\right)\left(\frac{1}{60}\right)$$

= 3051 cfm

humid heat = .24 + .45H = .24 + .45(.007)
= .243 Btu/°F lb air

$$Q = 1365 \frac{\text{lb air}}{\text{hr}}\left(.243 \frac{\text{Btu}}{°\text{F lb}}\right)(300-68) = 770040 \frac{\text{Btu}}{\text{hr}}$$

[8]

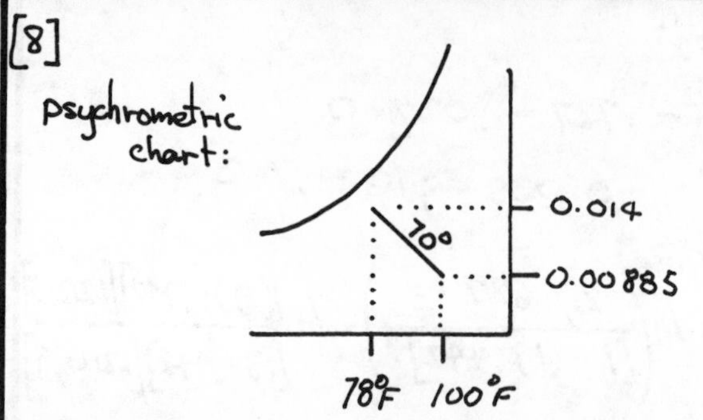

temperature of humidified air = 78°F

humidity of air to tower = 0.00885 $\frac{\text{lb } H_2O}{\text{lb air}}$

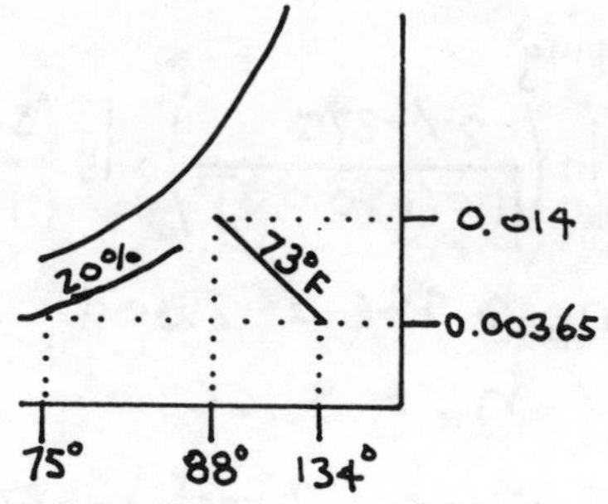

from psychro. chart:

water temp: 73°F

air temp leaving heater: 134°F

temp of humidified air: 88°F

H_{air} from preheater: 29.5 Btu/lb

H_{air} to preheater: 14.5 Btu/lb

heat added: 29.5 − 14.5 = 15 Btu/lb

[9]

$$\text{Initial free moisture} = 0.3 - 0.02 = 0.28 \frac{\text{lb } H_2O}{\text{lb d.s.}}$$

$$\text{Final free moisture} = 0.1 - 0.02 = 0.08 \frac{\text{lb } H_2O}{\text{lb d.s.}}$$

$$\text{drying time to } 10 \frac{\text{lb } H_2O}{\text{lb d.s.}} = 6 = K(.28 - .16) + K\,0.16 \ln\left(\frac{0.16}{0.08}\right)$$

K = 25.97

$$\Theta_T = K(0.12) + K \ln \frac{0.16}{0.04}$$

Θ_T = 8.86 hours

[10]

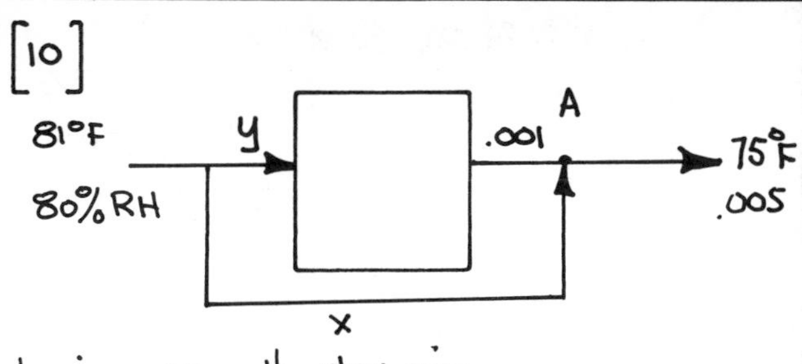

basis: 100 lb dry air

from psych. chart, air in: $0.018 \frac{lb H_2O}{lb\, d.a.}$

x, y = lb dry air bypassed, absorber

H_2O balance around point A

$$y(.001) + x(.018) = 100(.005)$$

$$y + x = 100$$

$$x = 25.53 \text{ lb d.a.}$$

% bypassed = 25.53%

[11]

2000 gpm
115°F

75°F
45% RH

2000 gpm
85°F

Assume Air leaves sat @ 105°F

$H_{in} = 0.005$ lb H_2O/lb d.a

$H_{out} = 0.05$ lb H_2O/lb d.a.

$$q_{WATER} = WC_p\Delta t = 2000(8.53)(1)(115-85)$$

$$= 5\times10^5 \text{ Btu/min}$$

ENTHALPY FROM TABLES

in: $h_{air} = [h_{174°} + h_{76°}]/2$

$$= \frac{17.778 + (0.45)19.88 + 18.259 + (0.45)21.31}{2}$$

$$= 27.286 \text{ Btu/lb d.a.}$$

out: $h_{air} = [h_{104°} + h_{106°}]/2$ (sat.)

$$= (79.31 + 83.42)/2$$

$$= 81.37 \text{ Btu/lb d.a.}$$

$$\Delta h_{tower} = 81.37 - 27.29 = 54.08 \text{ Btu/lb d.a}$$

$$\text{lb d.a./min} = 5\times10^5/54.8 = 9250 \text{ lb d.a./min}$$

Specific volume air @ 75° & 45% RH

$$= \frac{13.398 + 0.45(.364) + 13.449 + 0.45(.422)}{2}$$

$$= 13.6 \text{ ft}^3\text{/lb d.a.}$$

Volume = 13.6(9250) = 125803 ft³/min

make up H_2O: $\frac{(1.9)(.05-.005)9250}{8.33} = 94.9$ gpm

at the boiling point of an azeotrope, the vapor and liquid composition are the same.

[12] Air properties at 29.92 in Hg

$H_{in} = .0035$ lb H_2O/lb d.a. (w.b. = 30°)

$H_{out} = .3289$ lb H_2O/lb d.a. (w.b. = 162.5°)

correction for pressure, use tables in Perry:

in: $\Delta H = 0.6 - \frac{70}{24}(.01)(0.6) = .582 \frac{\text{grains } H_2O}{\text{lb d.a.}}$

$$= 0.000083 \frac{lb\, H_2O}{lb\, d.a.}$$

$$H = .0034 \frac{lb\, H_2O}{lb\, d.a} \text{ @ 14.3 psia}$$

out: $\Delta H = .3289 - \frac{180}{24}(.1)(56) = 51.8 \frac{\text{grains}}{\text{lb}}$

$$= .0074 \text{ lb/lb}$$

$$H = .3289 - .0074 = 0.3215 \text{ lb } H_2O\text{/lb d.a.}$$

evaporated: $.3215 - .0034 = .3181 \frac{lb\, H_2O}{lb\, d.a.}$

$$\text{lb d.a} = \frac{1}{.3181}\left(12 \frac{lb\, H_2O}{hr}\right) = 37.72 \frac{\text{lb d.a}}{\text{hr}}$$

$$\text{humid volume} = \frac{0.754(t+459.8)}{29.1}\left(1 + .0034[1.606]\right)$$

$$= 13.80 \text{ ft}^3\text{/lb d.a.}$$

$$Q = 37.72(13.80)/60 = 8.67 \text{ ft}^3\text{/min}$$

[13] see graphical solution 9-13

$$R_{D\,min} = \frac{x_D - y'}{y' - x'}$$ where x′, y′ are coordinates of intersection of feed line & equil line

$$R_D = \frac{0.95 - 0.70}{0.70 - 0.40} = 0.833$$

[13] con't

refer to graphical solution 9-13

For total reflux 4.6 plates + reboiler

Reflux ratio = $2(R_{D\,min}) = 1.667$

the y intercept in the rect. section is obtained from

$$y = \frac{R_D}{R_D+1}x + \frac{x_D}{R_D+1} \quad \{x=0\}$$

$$y' = \frac{x_D}{R_D+1} = \frac{.95}{2.667} = 0.356$$

Strip op. line drawn from intersection of rect. op. line with q line at $x_B = .05$

No of plates = 7.7 + reboiler

Actual plates = $\frac{7.7}{.6} = 12.8$ + reboiler

Feed line intersects 5th plate from top (= feed plate)

[14] refer to graphical solution 9-14

$$y = \frac{R}{R+1}x + \frac{x_D}{R+1} \quad @x=0;\ y = \frac{.97}{3+1} = .2425$$

op. line constructed from (.97, .97) to (0, .2425)

feed line (q line) is vertical

strip op. line drawn from q line intersection to (.04, .04)

theo. plates = 6.8 − 1 = 5.8 + reboiler

actual plates = $\frac{5.8}{.65} = 8.9$ + reboiler

feed plate on theo. 4.2 plate = 7 actual from top.

Use 12 plates for safety factor

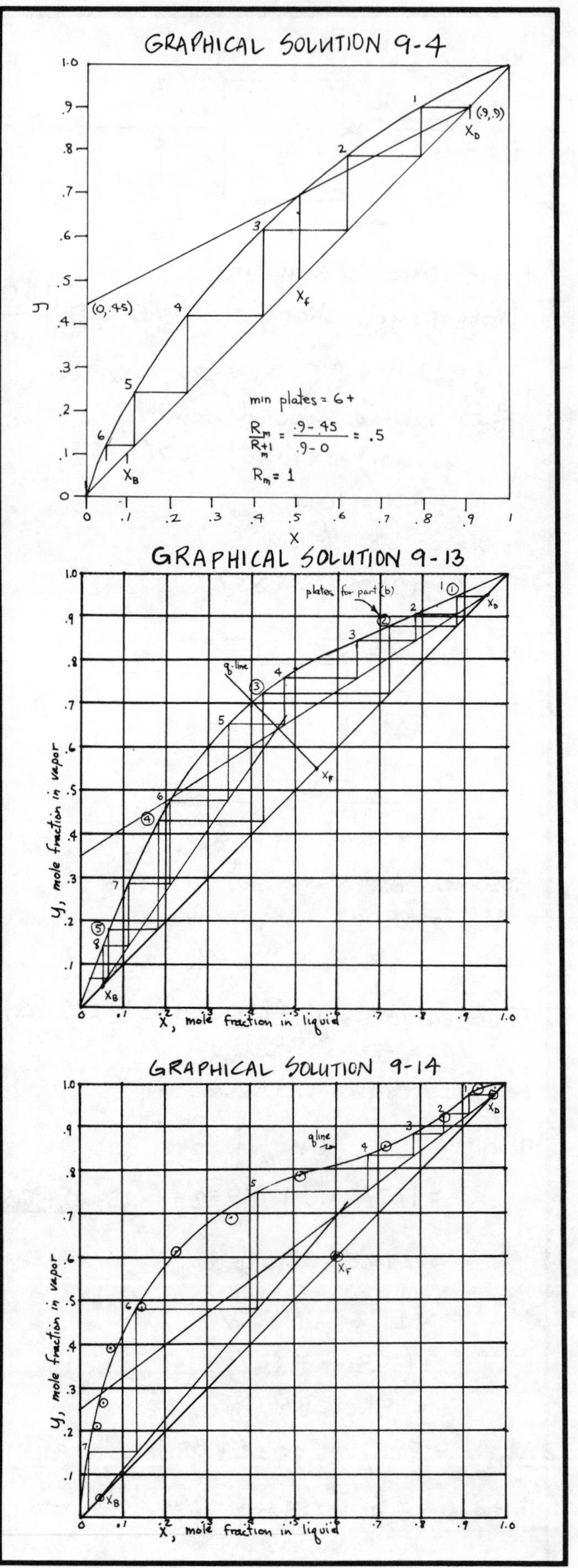

[1]

EXTRACT #1

	START	ADD	AFTER MIX	IN SOLN	DRY SOLIDS	DRAIN SOLN	DRAIN SOLIDS
SAND	200	–	200	–	200	–	200
KCl	800	50	850	447 [a]	403	193 [b]	596
H_2O	–	950	950	950	–	410	410
Σ	1000	1000	2000	1397	603	603	1206

a: H_2O can hold 950(.471) = 477 lb KCl

b: $KCl = \frac{603}{1397} \times 477 = 193$ lb

EXTRACT #2

	START	ADD	AFTER MIX	IN SOLN	DRY SOLIDS	DRAIN SOLN	DRAIN SOLIDS
SAND	200		200	–	200		200
KCl	596		596	596	– [a]	53.9 [b]	53.9
H_2O	410	1206	1616	1616	–	146.1	146.1
Σ	1206	1206	2412	2212	200	200	400

Dry:

	SOLIDS	wgt %
SAND	200	78.8
KCl	53.9	21.2
H_2O	–	
	253.9	100

a: H_2O in EXTR #2 CAN HOLD 1616(.471) = 761 lb KCl
∴ all dissolves

b: $KCl = \frac{200}{2212}(596) = 53.9$ lb

[2]

Extr-I	Extr #1	Extr #2	Σ	wgt %
KCl	254	542.1	796.1	28.4
H_2O	540	1469.9	2009.9	71.6
Σ	794	2012.0	2806.0	100.

[3]

if N is the amount of nicotine extracted into kerosene

$$\frac{N \text{ lb nicotine}}{150 \text{ lb kerosene}} = 0.91\,\frac{1-N \text{ lb nicotene}}{99 \text{ lb } H_2O}$$

$N = 0.5796$

$$\%\text{ extracted} = \frac{.5796}{1} \times 100 = 57.96\%$$

[4]

Basis: 1 lb A+B

$V_0 = \frac{\text{lb solv. feed}}{\text{lb (A+B)}}$

$$Y_1 = \frac{.3}{V_0 + .3}$$

90% of A in extract = 0.9(0.3) = 0.27 lb

A in underflow = 0.3 − .27 = 0.03 lb

solution in underflow =

$$0.7 \text{ lb B}\,\frac{\text{lb solution}}{\text{lb B}} = 0.7 \text{ lb solution}$$

solvent in underflow = 0.7 − .03 = 0.67 lb

$$Y_1 = \frac{0.03}{0.7} = 0.0428\,\frac{\text{lb A}}{\text{lb solution}} \quad \text{overflow}$$

$$Y_1 = 0.0428 = \frac{0.3}{V_0 + .03} = 6.7\,\frac{\text{lb solvent}}{\text{lb (A+B)}}$$

[5]

NaOH balance around stage 1:

$80 + W = 78.4 + .1(200)$

$W = 18.4$ lb in wash entering stage 1

Wash entering stage 1:

$$y_2 = 18.4/504 = 0.0365\,\frac{\text{lb NaOH}}{\text{lb solution}}$$

Overall NaOH balance

$80 + 0 = 78.4 + W_{\text{in underflow}}$

$W_{\text{in underflow}} = 1.6$

$$X_N = 1.6/200 = 0.008\,\frac{\text{lb NaOH}}{\text{lb solution}}$$

$X_0 = 80/480 = .1667$ lb NaOH/lb solution

$X_1 = 0.1$

$$n-1 = \frac{\ln \dfrac{y_{N+1} - X_n}{y_2 - X_1}}{\ln \dfrac{y_{N+1} - y_2}{X_n - X_1}} = \frac{\ln \dfrac{0 - .008}{.0365 - .1}}{\ln \dfrac{0 - .0365}{.008 - .1}} = 2.24$$

$n = 3.24$ stages

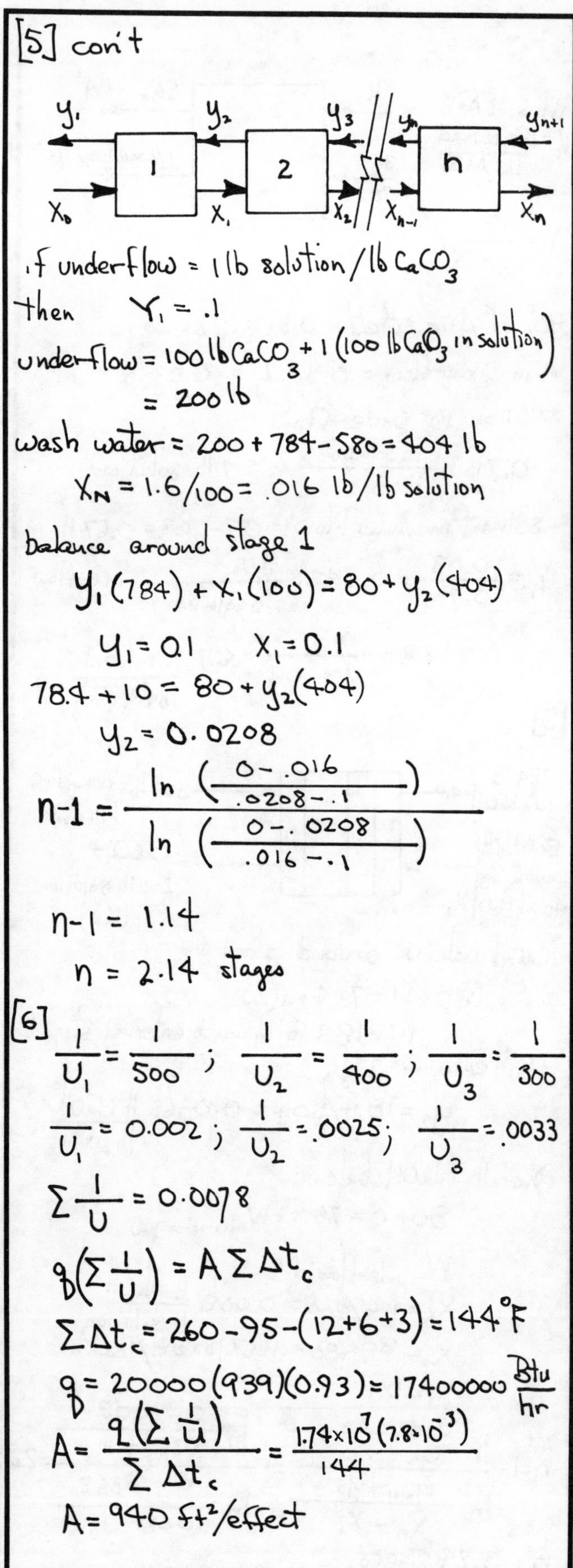

[5] con't

if underflow = 1 lb solution/lb $CaCO_3$

then $Y_1 = .1$

underflow = 100 lb $CaCO_3$ + 1 (100 lb $CaCO_3$ in solution)

= 200 lb

wash water = 200 + 784 − 580 = 404 lb

$X_N = 1.6/100 = .016$ lb/lb solution

balance around stage 1

$$y_1(784) + x_1(100) = 80 + y_2(404)$$

$y_1 = 0.1 \quad x_1 = 0.1$

$$78.4 + 10 = 80 + y_2(404)$$

$$y_2 = 0.0208$$

$$n-1 = \frac{\ln\left(\frac{0-.016}{.0208-.1}\right)}{\ln\left(\frac{0-.0208}{.016-.1}\right)}$$

$n - 1 = 1.14$

$n = 2.14$ stages

[6]

$$\frac{1}{U_1} = \frac{1}{500}; \quad \frac{1}{U_2} = \frac{1}{400}; \quad \frac{1}{U_3} = \frac{1}{300}$$

$$\frac{1}{U_1} = 0.002; \quad \frac{1}{U_2} = .0025; \quad \frac{1}{U_3} = .0033$$

$$\Sigma \frac{1}{U} = 0.0078$$

$$q\left(\Sigma \frac{1}{U}\right) = A \Sigma \Delta t_c$$

$$\Sigma \Delta t_c = 260 - 95 - (12 + 6 + 3) = 144°F$$

$$q = 20000(939)(0.93) = 17400000 \ \frac{Btu}{hr}$$

$$A = \frac{q\left(\Sigma \frac{1}{U}\right)}{\Sigma \Delta t_c} = \frac{1.74 \times 10^7 (7.8 \times 10^{-3})}{144}$$

$A = 940$ ft²/effect

$$\Delta t_1 = \frac{1/U_1}{\Sigma 1/U} \Sigma \Delta t_c$$

$$= \frac{0.002}{0.0078}(144) = 37°F$$

$$\Delta t_2 = \frac{.0025}{.0078}(144) = 46°F$$

$$\Delta t_3 = \frac{.0033}{.0078}(144) = 61°F$$

temperatures

Chest I = 260°F

Vapor I = Chest II = 232 − 12 = 211°F

Body I = 260 − 37 = 223°F

Chest III = Vapor II = 165 − 6 = 159°F

Body II = 211 − 46 = 165°F

Vapor III = Condenser = 98 − 3 = 95°F

Body III = 159 − 61 = 98°F

Vapor from each effect is approx

20000 lb/hr

$$60000 = 0.6 w_f \left(\frac{94}{6} - \frac{50}{50}\right)$$

$$w_f = \frac{60000}{0.6(14.8)} = 67500 \ \frac{lb}{hr}$$

[7]

W_1, W_2; F: 10% NaOH, 15.5 NaCl → 1 → F_2: 20% NaOH → 2 → B_2: 50% NaOH, 1% NaCl; B_1 → C → B_c

$$100 \text{ tons/day}\left(\frac{1}{22}\right) = 0.5 B_2$$

$B_2 = 18182$ lb/hr

balance around effect 2:

NaOH: $0.2 F_2 = 0.5 B_2$

NaCl $\quad X F_2 = .01 B_2$

$X = 0.004$ or 0.4 wgt% in m.l.

Composition of B_c:

$$\frac{5 \text{ lb } H_2O}{100 \text{ lb solid}} * \frac{0.2 \text{ lb NaOH}}{0.796 \text{ lb } H_2O} = 1.256 \ \frac{\text{lb NaOH}}{100 \text{ lb solid}}$$

[7] con't B_c: 5% H_2O, 1.256% NaOH, 93.74% NaCl

B_2:
NaOH: $0.5(18182) = 9091$
NaCl: $0.01(18182) = 181.8$
H_2O: $-(9091 + 181.8) + 18182 = 8909.2$

F_2
NaCl: $.004 F_2 = 181.8$
$F_2 = 45455$

Overall balance

$$F = W_1 + W_2 + B_c + B_2$$

NaOH
$$.105 F = 0.5(18182) + .01256 B_c$$
NaCl
$$.155 F = 0.01(18182) + .9374 B_c$$

$F = 107640$
$B_c = 17605$

material balance:
lb/hr:

	NaOH	NaCl	H_2O	Σ
F	11302	16684	79654	107640
B_1	11302	16684	37062	65048
W_1	–	–	42592	42592
B_c	2211	16502	880	17605
F_2	9091	181.8	36182	45455
W_2	–	–	27273	27273
B_2	9091	181.8	8909	18182

[8] heat:
evaporation: $(14000)(970) = 1.36\times10^7 \frac{\text{Btu}}{\text{hr}}$
heating: $(8.33)(60)(100)(1.05)(1)(212-185)$
$= 1.42\times10^6$ Btu/hr

heat req'd = 1.502×10^7 Btu/hr
let x = lb/hr of 120 psia steam
x = lb/hr of vapor compressed from 212° to 233°

$H_{in} = 1190.4(x)$ 120 psi steam
$H = 1150.4(x)$ 212°F vapor

$h_{condensate} = (2x)(201.26)$

H balance
$$1190.4(x) + 1150.4x - (2x)(201.26) = 1.502\times10^7$$

x = 7750 lb/hr steam

Other types of recompression are mechanical: centrifugal

[9] $\mu = 0.982 \times 6.72\times10^{-4} = 6.6\times10^{-4} \frac{\text{lb}}{\text{ft·sec}}$

$$K = D_p\left[\frac{g(\rho_p - \rho)}{\mu^2}\right]^{1/3}$$

stokes law: $$V_t = \frac{g D_p^2(\rho_p - \rho)}{18\mu}$$

intermediate law: $$V_t = \frac{0.153\, g^{0.71} D_p^{1.14}(\rho_p - \rho)^{.71}}{\rho^{0.29}\mu^{0.43}}$$

for galena
$$K = D_p\left[\frac{(32.17)(62.3)(6.5)(62.4)}{(6.6\times10^{-4})^2}\right]^{\frac{1}{3}} = D_p\left[1.231\times10^4\right]$$

since $D_p = .01$ K = 10.5 ∴ obeys intermediate law

$$v_t = \frac{0.153(32.17)^{.71}\left(\frac{.01}{12}\right)^{1.14}\left[6.5(62.4)\right]^{.71}}{(62.3)^{.29}(6.6\times10^{-4})^{0.43}}$$

$v_t = .279$ ft/sec
$t = 5/.279 = 18$ sec.

.001" diameter K = 1.025 < 3.5 (stokes)
$$v_t = \frac{(32.17)(.833)^2\, 10^{-8}(406)}{18(6.6\times10^{-4})} = 0.00761 \frac{\text{ft}}{\text{sec}}$$

$t = 5/.00761 = 657$ sec

for quartz
similarly $D_p = 0.01$ K = 6.5 Intermediate
$v_t = 0.1049$ ft/sec ; t = 48 sec

$D_p = .001$ Stokes
$v_t = .001932$ ft/sec ; t = 43 min

[10]

$$Q_c = \frac{(\rho_s - \rho) D_p^2 V \omega^2 r}{9 \mu s}$$

$$V = \frac{16}{12}\pi\frac{(D_1^2 - D_2^2)}{4} = \frac{\pi}{3}(2^2 - 1.5^2) = 1.83 \text{ ft}^3$$

$\mu = 3 \text{ cp} = 2.016 \times 10^{-3}$ lb/ft-sec

$(\rho_s - \rho) = (1.6 - 1.3)(62.4) = 18.72$ lb/ft^3

$\omega = 1000(2\pi)/60 = 104.7 \text{ sec}^{-1}$

$r = 1$ ft

$s = 0.25$ ft

$D_p = 30 \mu m = 9.84 \times 10^{-5}$ ft

$$Q_c = \frac{(18.72)(9.84 \times 10^{-5})^2(1.83)(104.7)^2(1)}{9(2.016 \times 10^{-3})(.25)}$$

$Q_c = 0.8016$ ft^3/sec $= 359.8$ gpm

[12]

$$\text{rate} = \sqrt{\frac{2 \Delta P f}{\mu \alpha \omega \theta_\pi}}$$

assume at new rate ΔP, f, and properties of cake constant

$$\frac{r_2}{r_1} = \sqrt{\frac{\theta_{\pi\,1}}{\theta_{\pi\,2}}}$$

doubling rpm, halves cycle time

$$\frac{r_2}{r_1} = \sqrt{\frac{1}{(1/2)}} = 1.414$$

$r_2 = r_1(1.414)$ or 41.4% increase

[1]

$$\ln\frac{10}{1} = \frac{E}{R}\left[\frac{1}{400} - \frac{1}{500}\right]; \quad R = 1.987$$

$$\frac{E}{R} = 4605.1; \quad E = 9.15 \text{ kcal/mole}$$

$$\ln\frac{K_2}{K_1} = \frac{E}{R}\left[\frac{1}{400} - \frac{1}{450}\right]$$

$$= 4605.1\left[\frac{1}{400} - \frac{1}{450}\right] = 1.27919$$

$\frac{K_2}{K_1} = 3.59$ times as fast at 450°K

[2]

$$\frac{dC_A}{dt} = kC_A^{0.5} \qquad C_A = C_{A_0}(1 - X_A)$$

$$kt = 2C_{A_0}^{0.5}\left[1 - (1 - X_A)^{0.5}\right]$$

@ $t = 0$; $X_A = 0.75$; $k = C_{A_0}^{0.5}/10$

@ $t = 30$

$$\frac{C_{A_0}^{1/2}}{10}(30) = C_{A_0}^{1/2}\left[1 - (1 - X_A)^{.5}\right];$$

$X_A = 0.75$ (not possible)

since at $t = 20$

$$\frac{C_{A_0}^{1/2}}{10}(20) = 2C_{A_0}^{1/2}\left[1 - (1 - X_A)^{1/2}\right]$$

$$X_A = 1.00$$

∴ @ $t = 30$ $X_A = 1.00$ also

[3]

$$\ln\frac{1}{1 - X_A} = kt; \quad \ln\frac{1}{1 - .5} = \ln 2 = k(5)$$

$$k = \ln 2/5$$

at $X_A = .75$

$$\ln\frac{1}{1 - .75} = \ln 4 = kt = \frac{\ln 2}{5}t$$

$$t = 5\frac{\ln 4}{\ln 2} = (5)\frac{2\ln 2}{\ln 2} = 10 \text{ minutes}$$

[4] Ideal gas

$$\frac{P_1}{T_1} = \frac{P_2}{T_2}; \quad P_2 = 1\left(\frac{373}{298}\right) = 1.25 \text{ atm}$$

$$-\frac{dC_A}{dt} = kC_A^2; \quad C_A = \frac{P_A}{RT}$$

$$RT\left(\frac{1}{P_A} - \frac{1}{1.25}\right) = kt \qquad R = 0.0821 \frac{\text{liter-atm}}{\text{mole °K}}$$

$2A \rightarrow B \qquad \Delta n = 1 - 2 = -1$

$$P_A = P_{A_0} - \frac{2}{-1}(\pi - \pi_0); \quad P_A = 2\pi - 1.25$$

t	π	P_A	kt
0	1.25	1.25	0
1	1.14	1.03	5.23
2	1.04	.83	12.40
3	0.982	.714	18.39
4	0.940	.63	24.11
5	0.905	.56	30.19
6	0.870	.49	38.00
7	0.850	.45	43.55
8	0.832	.414	49.47
9	0.815	.38	56.09
10	0.800	.35	63.00
15	0.754	.258	94.20
20	0.728	.206	124.16

$\Sigma = 90 \qquad\qquad \Sigma = 558.79$

$$K = \frac{558.79}{90} = 6.2088$$

rate equation at 373°K

$$\frac{1}{P_A} = 0.203t - 0.8$$

P_A is partial pressure A in atm

t is time, minutes

[5] BASIS: 1 lbmole chlorohydrin

	lbmoles	lb	
chlorohydrin	1.0	80.52	FEED 414.44
water		241.5	
bicarbonate	1.10	92.42	
glycol	.99		PRODUCT 370.9 lb
NaCl	.99		
$NaHCO_3$	.11		
Chlorohydrin	.01		
Water	–	241.5	
$CO_2\uparrow$	.99	43.6	

$$V = \frac{370.9}{8.33(1.1)} = 40.4 \text{ gal}$$

$X_A = 0.99$; $K_c = 1250 \frac{\text{gal}}{\text{unit charge/hr}}$

$n_A = .01$ $n_B = 0.11$

$$\frac{V_2}{F} = \frac{X_A}{K_c \frac{n_A}{V}\frac{n_B}{V}} = \frac{0.99}{1250 \frac{.01}{40.4}\frac{.11}{40.4}}$$

$$\frac{V_2}{F} = 1175 \frac{\text{gal}}{\text{unit charge/hr}}$$

1 unit charge = 61.5 lb product

$$F = \frac{10}{61.5} = 0.1625 \text{ unit charge/hr}$$

$$V_2 = 0.1625(1175) = 191 \text{ gal}$$

[6] $(CH_3CO)_2O + H_2O \rightarrow 2CH_3COOH$

$$\frac{1}{V}\frac{dN_a}{dt} = -kC_{H_2O}C_A$$

(a) since solution is dilute, H_2O concentration constant so kC_{H_2O} is constant

(b) integrated 2nd order batch

$$\ln[(1-X_B)/(1-X_A)] = (C_{B_0} - C_{A_0})kt$$

$C_B \approx C_{B_0}$ $X_B \approx 0$ $C_{B_0} >> C_{A_0}$

therefore

$$-\ln(1-X_A) = C_{B_0}kt = k't$$

$$-\ln(1-.7) = 0.0806t$$

(c) $k = k_0 e^{-E/RT}$ or $kC_{H_2O} = k_0' e^{-E/RT}$

$T = 283.15$ $k_0' e^{-E/R(283.15)} = 0.0567$

$T = 313.15$ $k_0' e^{-E/R(313.15)} = 0.380$

$$\frac{E}{R}\left(\frac{1}{283.15} - \frac{1}{313.15}\right) = 1.9$$

$\frac{E}{R} = 5622$ °K @ 313.15 $k_0' = .0567 e^{\frac{5622}{313.15}}$

$$k_0' = 3.55\times10^6$$

$$-r_A = C_A(3.55\times10^6)e^{-5622/T}$$

[7] $O_2 + 2NO \rightarrow 2NO_2$ (A) = O_2, (B) = NO

Component	initially moles	moles@90%
O_2	.1053	$.1053 - .9\frac{.1842}{2} = .02241$
NO	.1842	$(.1)(.1842) = .01842$
N_2	.7105	.7105
NO_2	0	$.9(.1842) = .16578$
	1.0000	.91711

Vol. of 1 gmole gas @ 1 atm 30°C

$$V_R = 22.4(1)(303.15/273.15)$$

$$V_R = 24.86 \text{ liters}$$

Integrated 3rd order equation

$$\frac{(2C_{A_0} - C_{B_0})(C_{B_0} - C_B)}{C_{B_0}C_B} + \ln\frac{C_{A_0}C_B}{C_A C_{B_0}} = (2C_{A_0} - C_{B_0})^2 kt$$

$C_i = n_i/V$ volume terms cancel left side

$$\frac{2(.1053) - .1842(.1842 - .01842)}{.1842(.01842)} + \ln\frac{(.1053)(.01842)}{(.02241)(.1842)} = \frac{(2(.1053) - .1842)^2}{V^2}kt$$

$$t = 17.89 \text{ sec}$$

[7] con't

i		X_i
O_2	.02241/.91711 =	.0244
NO	.01842/.91711 =	.0201
N_2	.7105/.91711 =	.7747
NO_2	.16578/.91711 =	.1808

total pressure proportional to moles present

$p = .91711(760) = 697$ mmHg

[8] total inlet feed = 9 gpm

$C_{A_0} = (3)(5)/9 = 1.667 \frac{\text{lbmole}}{\text{gal}}$

$C_{B_0} = (1.5)(4)/9 = 0.667 \frac{\text{lbmole}}{\text{gal}}$

80% conversion $C_B = 0.2(.667) = .1334$

$C_A = 1.667 - [.667 - .1334] = 1.133$

$\frac{V}{q} = \frac{.667 - .133}{(1)(1.133)(.1334)}$ $q = 9$ gpm

$V = 31.8$ gal.

[9] Arrhenius plot

$\frac{d \ln k}{dT} = \frac{E}{RT^2}$

$K = K_0 \, 2^{(T_2 - T_1)/10}$

$\ln k = \ln K_0 + \frac{T_2 - T_1}{10} \ln 2$

$\frac{d \ln k}{dT} = \frac{\ln 2}{10} = .0695 = \frac{E}{RT^2}$

$E = 1.383 \times 10^{-4} T^2$

T °K	300	400	600	800	1000
E	12.45	22.1	49.8	88.5	138.3

[10] assume first order

$\ln \frac{C_A}{C_{A_0}} = -kt$

at 100°C; 120°C; 150°C

C_A	C_A/C_{A_0}	$-\ln C_A/C_{A_0}$	t (100°C)	t (120°C)	t (150°C)
15	1	0	0	0	0
12	.8	0.2231	111.	55.3	22.7
10.5	.7	0.3567	178		
7.5	.5	0.6931	346	172	70.5
6	.4	0.9162	457		
4.5	.3	1.2040	600	300	122.4
3.0	.2	1.6094	804	402.2	164.1
1.5	.1	2.3026	1145	573.1	232.0
		Σ = 7.3051	Σ = 3641		
		Σ = 6.0322		Σ = 1502.6	
		Σ = 6.0322			Σ = 611.7

plot $-\ln \frac{C_A}{C_{A_0}}$ versus t

yields straight line ∴ 1st order

$K_{100} = 7.3051/3641 = 2.006 \times 10^{-3}$

$K_{120} = 6.0322/1502.6 = 4.0145 \times 10^{-3}$

$K_{150} = 6.0322/611.7 = 9.861 \times 10^{-3}$

$\ln \frac{K_1}{K_2} = -\frac{E}{R}\left[\frac{1}{T_1} - \frac{1}{T_2}\right]; \; -\frac{E}{R} = \frac{\ln K_1/K_2}{\frac{1}{T_1} - \frac{1}{T_2}}$

$\ln \frac{K_1}{K_2} / \left[\frac{1}{T_1} - \frac{1}{T_2}\right] = -5095$

$\ln \frac{K_1}{K_3} / \left[\frac{1}{T_1} - \frac{1}{T_3}\right] = -5019$

$\ln K_2/K_3 / \left[\frac{1}{T_2} - \frac{1}{T_3}\right] = -4963$

avg = −5026

since $K = K_0 e^{-E/RT}$ @ T = 373, K = 2.006, $-E/R = -5026$

$K_0 = Ke^{E/RT} = 2.006 e^{5026/373} = 1.426 \times 10^6$

∴ $K = 1.426 \times 10^6 e^{-5026/T}$

[11] 1st order

$$\ln \frac{E}{E_o} = -kt \; ; \; K = \frac{1}{t} \ln \frac{E_o}{E}$$

$$K_{70} = \frac{1}{10} \ln \left[\frac{.175}{.175 - \frac{.022}{2}} \right] = 6.492 \times 10^{-3} \text{ min}^{-1}$$

$$K_{100} = \frac{1}{10} \ln \left[\frac{.175}{.175 - \frac{.059}{2}} \right] = 1.846 \times 10^{-2} \text{ min}^{-1}$$

$$\ln \frac{K_{100}}{K_{70}} = \frac{E}{R} \left[\frac{1}{530} - \frac{1}{560} \right] = 1.045$$

$$\frac{E}{R} = 10340$$

@ 170°F (630°R)

$$\ln \frac{K_{130}}{K_{70}} = 10340 \left[\frac{1}{530} - \frac{1}{630} \right] = 3.097$$

$$K_{130} = .1436$$

80% conversion

$$\ln \frac{E_o}{E} = \ln \frac{E_o}{.2E_o} = \ln 5 = 0.1436t$$

$$t = 11.21 \text{ minutes}$$

[12] ½ life = 1.3 hr

$$\therefore \ln \tfrac{1}{2} = -K(1.3); \; K = 0.532 \text{ hr}^{-1}$$

$$\ln \frac{A}{A_o} = \ln 0.30 = -kt; \; t = 2.26 \text{ hr}$$

$$\frac{200 \text{ liters}}{2.26 \text{ hr}} = 88.5 \text{ liters/hr}$$

In CSTR

$$qA_o = qA + KVA; \; A = 0.3A_o$$

$$q = .3q + .3KV$$

$$q = \frac{.3}{.7} KV = \frac{.3}{.7}(400)(.532)$$

$$q = 91.0 \text{ liters/hr}$$

Choose CSTR

[13] $2P \rightarrow Q + R$

$$\frac{1}{C_P} - \frac{1}{C_{P_o}} = kt = (0.2)(1)$$

$$\frac{1}{C_P} = 0.8666$$

$$C_P = 1.1538$$

or $1.5 - 1.1538 = 0.346$ moles/liter of P reacted

$\therefore$ $0.346/2$ = moles/liter Q formed $= 0.173$

$$\text{Volume} = \frac{100}{.173} = 578 \text{ liters}$$

let q = vol. feedrate

$$qP_o - 2VkP^2 = qP$$

200 moles P must disappear

$$q(P - P_o) = 200 = 2(356)(.2)(P)^2$$

$P = 1.185$ yes, since P is less than 1.5

$$q = \frac{200}{1.5 - 1.185} = 635 \text{ liters/hr}$$

[14] $\tau = \frac{V}{v_o} = \frac{C_{A_o} X_A}{-r_A}$ $C_{A_o} = C_{B_o}$

$$\frac{dC}{dt} = k C_A C_B = k C_{A_o}(1 - X_A) C_{B_o}(1 - X_B)$$

$$= k C_{A_o}^2 (1 - X_A)^2$$

$$-r_A = -r_B = r_o$$

$$\tau = \frac{C_{A_o} X_A}{-r_A}; \; \tau = \frac{2000}{6000} = .33 \text{ hr}$$

[14] con't

$$\tau = .33\text{hr} = 20\text{min} = \frac{C_{A_0} X_A}{K C_{A_0}^2 (1-X_A)^2}$$

$$= \frac{X_A}{17.1(.2)(1-X_A)}$$

solving quadratic

$$X_A = 0.886$$

$$C_D = (.886)(.2) = 0.177 \frac{\text{lbmoles}}{\text{ft}^3}$$

[15] 2A → products (2R+S)

variable volume, batch

$$t K C_{A_0} = \frac{(1+\varepsilon_A) X_A}{1-X_A} + \varepsilon_A \ln(1-X_A)$$

$$\varepsilon_A = \frac{3-2}{2} = \frac{1}{2}$$

$$C_{A_0} = \frac{n_{A_0}}{V_{A_0}} = \frac{P}{RT} = \frac{1\text{ atm}}{0.0821(273+950)}$$

$$C_{A_0} = 9.959 \times 10^{-3} \text{ gmoles/liter}$$

$$K = 1200 \text{ cc/gmole}\cdot\text{sec} = 1.2 \frac{\text{liter}}{\text{gmol}\cdot\text{sec}}$$

@90% $X_A = .9$

$$t(9.959 \times 10^{-3})(1.2) = \frac{(1+\frac{1}{2}).9}{(1-.9)} + \frac{1}{2}\ln(1-.9)$$

$$t = 1033 \text{ sec}$$

Quick – I need additional study materials!

Please rush me the review materials I have checked. I understand any item may be returned for a full refund within 30 days. I have provided my bank card number as method of payment, and I authorize you to charge your current prices against my account.

	Solutions Manuals:
For the E-I-T Exam:	
[ ] Engineer-In-Training Review Manual	[ ]
[ ] E-I-T Quick Reference Cards	
[ ] E-I-T Mini-Exam with Solutions	
For the P.E. Exams:	
[ ] Civil Engineering Reference Manual	[ ]
[ ] Civil Engineering Sample Exam with solutions	
[ ] Civil Engineering Quick Reference Cards	
[ ] Seismic Design	
[ ] Timber Design	
[ ] Structural Engineering Practice Problem Manual	
[ ] Mechanical Engineering Review Manual	[ ]
[ ] Mechanical Engineering Quick Reference Cards	
[ ] Electrical Engineering Review Manual	[ ]
[ ] Chemical Engineering Review Manual	
[ ] Chemical Engineering Practice Exam Set	
[ ] Land Surveyor Reference Manual	[ ]
Recommended for all Exams:	
[ ] Expanded Interest Tables	
[ ] Engineering Law, Ethics, and Liability	

SHIP TO:

Name ______________________

Company ______________________

Street ______________________ Apt. No. ________

City ______________________ State ______ Zip ______

Daytime phone number ______________________

CHARGE TO (required for immediate processing):

______________________ ________

VISA/MC/AMEX account number expiration date

____

name on card

____

signature

Send more information

Please send me descriptions and prices of all available E-I-T and P.E. review books. I understand there will be no obligation.

____

____

____

____

A friend of mine is taking the exam too. Send additional literature to:

____

____

____

____

I disagree...

I think there is an error on page ________. Here is the way I think it should be.

Title of this book: ______________________

[ ] Please tell me if I am correct.

Contributed by (optional):

____

____

____

Affix
Postage

PROFESSIONAL PUBLICATIONS, INC.
1250 Fifth Avenue
Belmont, CA 94002

Affix
Postage

PROFESSIONAL PUBLICATIONS, INC.
1250 Fifth Avenue
Belmont, CA 94002

Affix
Postage

PROFESSIONAL PUBLICATIONS, INC.
1250 Fifth Avenue
Belmont, CA 94002